Electronically Active Textiles

Electronically Active Textiles

Special Issue Editor
Tilak Dias

MDPI • Basel • Beijing • Wuhan • Barcelona • Belgrade

Special Issue Editor
Tilak Dias
Nottingham Trent University
UK

Editorial Office
MDPI
St. Alban-Anlage 66
4052 Basel, Switzerland

This is a reprint of articles from the Special Issue published online in the open access journal *Fibers* (ISSN 2079-6439) from 2018 to 2019 (available at: https://www.mdpi.com/journal/fibers/special_issues/electronically_active_textiles).

For citation purposes, cite each article independently as indicated on the article page online and as indicated below:

LastName, A.A.; LastName, B.B.; LastName, C.C. Article Title. *Journal Name* **Year**, *Article Number*, Page Range.

ISBN 978-3-03928-144-2 (Pbk)
ISBN 978-3-03928-145-9 (PDF)

Cover image courtesy of Tilak Dias.

© 2020 by the authors. Articles in this book are Open Access and distributed under the Creative Commons Attribution (CC BY) license, which allows users to download, copy and build upon published articles, as long as the author and publisher are properly credited, which ensures maximum dissemination and a wider impact of our publications.

The book as a whole is distributed by MDPI under the terms and conditions of the Creative Commons license CC BY-NC-ND.

Contents

About the Special Issue Editor . . . . **vii**

Preface to "Electronically Active Textiles" . . . . **ix**

Theodore Hughes-Riley, Tilak Dias and Colin Cork
A Historical Review of the Development of Electronic Textiles
Reprinted from: *Fibers* **2018**, *6*, 34, doi:10.3390/fib6020034 . . . . **1**

Alyssa Martin and Adam Fontecchio
Effect of Fabric Integration on the Physical and Optical Performance of Electroluminescent Fibers for Lighted Textile Applications
Reprinted from: *Fibers* **2018**, *6*, 50, doi:10.3390/fib6030050 . . . . **16**

Mohamad-Nour Nashed, Dorothy Anne Hardy, Theodore Hughes-Riley and Tilak Dias
A Novel Method for Embedding Semiconductor Dies within Textile Yarn to Create Electronic Textiles
Reprinted from: *Fibers* **2019**, *7*, 12, doi:10.3390/fib7020012 . . . . **29**

Dorothy A. Hardy, Andrea Moneta, Viktorija Sakalyte, Lauren Connolly, Arash Shahidi and Theodore Hughes-Riley
Engineering a Costume for Performance Using Illuminated LED-Yarns
Reprinted from: *Fibers* **2018**, *6*, 35, doi:10.3390/fib6020035 . . . . **46**

Pasindu Lugoda, Theodore Hughes-Riley, Carlos Oliveira, Rob Morris and Tilak Dias
Developing Novel Temperature Sensing Garments for Health Monitoring Applications
Reprinted from: *Fibers* **2018**, *6*, 46, doi:10.3390/fib6030046 . . . . **58**

About the Special Issue Editor

Tilak Dias obtained a 'Diplom-Ingenieur' in textile engineering from the Technische Universität in Dresden in 1981 after studying electrical and electronics engineering in 1969. He also obtained a 'Dr.-Ingenieur' from the Universität Stuttgart in 1988. Professor Dias has held the Chair in Knitting at Nottingham Trent University (United Kingdom) since 2010, where he is also the Head of the Advanced Textiles Research Group. He has over 38 years of experience in knitted structure design and knitting technology, and 17 years of experience in developing electronic textiles. Tilak has been invited to give over 60 talks and has published over 160 scientific papers in leading journals and conferences. His research focuses on the development of electronic textiles and advanced knitting technologies.

Preface to "Electronically Active Textiles"

Since their invention, textile materials have gone through many evolutions. Initially, the focus was on enhancing aesthetic properties (such as their colour, handle, and comfort). In the last century, the focus moved to improving textile functionality. This has led to the development of fabrics capable of stopping a bullet travelling at supersonic speeds, of fire retardant fabrics, and of impact- and cut-resistant fabrics. All these functionalities have been achieved using chemical processes and through advances in polymer science.

Textiles are now going through a new evolution to integrate electrical systems and electronic devices. Textiles are used to clothe our bodies because they are strong, soft, breathable, flexible, and conformable. The introduction of electronic components has the potential to compromise some of these highly-desirable characteristics; however, the proper integration would result in introducing intelligence to textile materials for the first time.

This book contains five peer-reviewed articles that review the development of electronic textiles (e-textiles) and original research articles. Topics include the fabrication and testing of e-textiles, the illumination of e-textiles, and e-textiles that can monitor skin temperature. This volume will be useful as a reference for designers and engineers in both academia and industry who would like to develop an understanding of e-textiles, as well as to students.

Tilak Dias
Special Issue Editor

Review

A Historical Review of the Development of Electronic Textiles

Theodore Hughes-Riley *, Tilak Dias and Colin Cork

Nottingham Trent University, Advanced Textiles Research Group, School of Art & Design, Bonington Building, Dryden St, Nottingham NG1 4GG, UK; tilak.dias@ntu.ac.uk (T.D.); colinrcork@googlemail.com (C.C.)
* Correspondence: theodore.hughesriley@ntu.ac.uk; Tel.: +44-(0)-11-5941-8418

Received: 13 February 2018; Accepted: 23 May 2018; Published: 31 May 2018

Abstract: Textiles have been at the heart of human technological progress for thousands of years, with textile developments closely tied to key inventions that have shaped societies. The relatively recent invention of electronic textiles is set to push boundaries again and has already opened up the potential for garments relevant to defense, sports, medicine, and health monitoring. The aim of this review is to provide an overview of the key innovative pathways in the development of electronic textiles to date using sources available in the public domain regarding electronic textiles (E-textiles); this includes academic literature, commercialized products, and published patents. The literature shows that electronics can be integrated into textiles, where integration is achieved by either attaching the electronics onto the surface of a textile, electronics are added at the textile manufacturing stage, or electronics are incorporated at the yarn stage. Methods of integration can have an influence on the textiles properties such as the drapability of the textile.

Keywords: electronic textiles; E-textiles; smart textiles; intelligent textiles

1. Introduction

This historical review will provide the reader with an insight into the development and employment of electronic textiles. While some academic sources have been used, a strong focus has been placed upon patented technology and commercialized products, which are often neglected in reviews of the literature.

The innovation of textiles 27,000 years ago could be contested as humanity's invention of the first material [1]. The passing of the millennia has consolidated humanity's need of textiles either to be protected from the environment, or desire to outwardly convey a message about themselves; whether that be artistic, stylistic, or wealth-related. The creation of textiles has therefore been coupled closely with key inventions that have shaped society; the knitting frame by William Lee in 1589 [2], the flying shuttle by John Kay in 1733 and the spinning jenny by James Hargreaves around 1765 [3], and set the foundation for the first industrial revolution.

A new revolution is now underway where the most widely used material by humankind has gained new functionality with the incorporation of electronic components. The first examples of electronic textiles date back to the use of illuminated headbands in the ballet La Farandole in 1883 [4]. More recently advances have appeared due to the reducing size and cost of electronic components, as well as an increased complexity of small-scale electronics, and have begun to show the true scope of possibilities for integrating electronics with clothing.

The growth of E-textiles in the later part of the 20th Century was due to a series of developments in material science and electronics further expanding the potential scope for embedding electronics within clothing. The conductive polymer was a key innovation; invented by Heeger et al. in 1977 [5] this led to a Nobel Prize thirty-three years later [6]. A patent for this type of technology for use with textiles was granted shortly after its creation [7]. Another critical development were advancements in transistor technology, with the creation of the first MOS (metal–oxide–semiconductor field-effect transistor) in 1960 [8]. The use of transistor-based electronics were outlined in a patent describing illuminated clothing from 1979 [9].

For a greater adoption of E-textiles a better level of integration of the electronic components was required. Key patents from 2005, 2016 and 2017 described the encapsulation of semi-conductor devices within the fibers of yarns [10–12]. This represented the start of the work on electronically functional yarns.

Three different pathways have been used to integrate electronics into textiles. These three distinct generations of electronic textiles are adding electronics or circuitry to a garment (first generation), functional fabrics such as sensors and switches (second generation), and functional yarns (third generation). Prior to the creation of E-textiles there are also many examples of the use of conductive fibers in textile fabrication, going back as far as the second century [13]. A timeline showing the evolution of E-textiles is given by Figure 1.

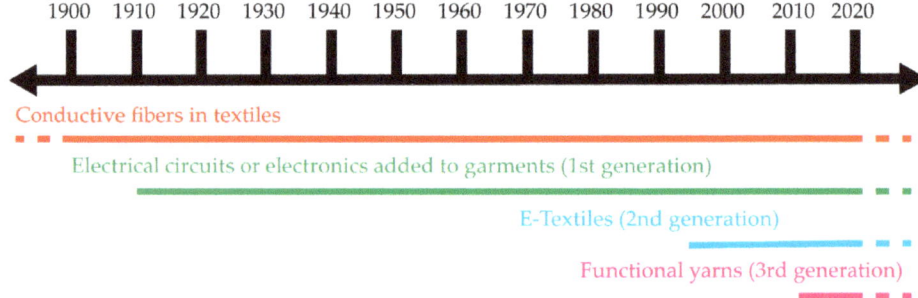

Figure 1. A timeline of the different generations of electronic textiles. This timeline shows when significant interest in the technology began, and not earlier, isolated instances.

Each method of integration will have an influence on the textile properties such as the shear properties of the textile, or its flexibility, both of which effect the drapability. Figure 2 shows examples of each generation of electronic textiles.

Figure 2. Photographs showing contemporary examples of each generation of electronic textile. (**Left**) An Adafruit coin cell battery holder. The first generation saw devices affixed to textiles. (**Middle**) A knitted electrode. The second generation of electronic textiles describes functional fabrics where conductive elements are integrated into a textile. (**Right**) An example of functional yarns (in this case LED yarns). The third generation of electronic textiles describe electronics embedded into textiles at a yarn level.

For the purposes of this review it is important to disentangle the various terms used loosely in the field of advanced textiles. Electronic textiles will be discussed; the strict definition of electronic textiles are where electronically conductive fibers or components are incorporated into a textile (electronic textiles will be referred to as 'E-textiles' in the subsequent text). Here, the term smart textiles, often anachronistically used as a synonym for electronic or E-textiles, will not be employed unless the textile has some kind of intelligence.

Critically, this historical review of the literature has a strong focus on innovation including patented technology and commercialized products, making heavy use of patents and internet sources. Non-internet sources have been referenced wherever possible; however, in many cases, particularly for products, only websites exist. These are areas that are typically neglected and this review serves as a complimentary piece to other more traditional reviews with the academic literature as the focus.

As a historic review this work will principally focus on research prior to 2010. Interest in electronic textiles has increased significantly since 2010. A search of the literature using the term 'electronic textiles' yields 310,000 results of which 254,000 are between 2010 and 2018. For completeness some more contemporary resources will be examined, for example in the area of textile energy solutions, which has largely evolved since 2010. There are several excellent reviews covering more recent developments in electronic textiles [14–16] and recent books on this subject [17,18].

The aim of this review is to overview the key pathways to the development of electronic textiles. The review is structured over ten sections, with the seven sections covering examples from a specific types of E-textile or application, such as sensors or illumination. Each section covers a major area of research in the literature, for example, temperature control textiles. In some cases two areas of interest have been grouped together such as E-textile pressure sensors and textile switches as one of these technologies led to the other. Three final sections discuss market trends, possible future developments for E-textiles, and a brief conclusion.

2. Temperature Control Textiles

It is unsurprising that the use of embedded conductive fibers was employed in one of the first true E-textiles, the electrically heated glove patented in 1910 [19]. This invention used ohmic heating (Joule heating, or resistive heating [20]), where an electric current was run through the conductive fibers and the electric resistance in the wires led to a heating effect. This type of heating was the core application of many early E-textiles. Other patents refining the heated gloves were filed [21,22], and variations on the theme appeared such as the heated boot [23], and full heated garments including a jacket element [24,25]. The knitted heater was patented in 1910 [26]. Heated gloves and jackets based on these core concepts are still available today showing the relevance of these early innovations.

Other devices utilizing heating elements were created between the 1930s and 1970s included the heated blanket [27], a heated baby carriage blanket [28], and electrically heated socks [29]. The year 1968 is often seen as the birth of modern E-textiles, when the Museum of Contemporary Craft in New York City held an exhibition, Body Covering, exploring a series of electric garments with functions such as heating and cooling [30].

Interest in developing new heated textiles remained well into the 21st Century with a patent from 2008 describing the incorporation of inductive heating elements into footwear and apparel [31]. Commercially, EXO Technologies have produced heated gloves that can be used by militaries or for a variety of outdoor activities such as skiing or motorcycling [32]. Marktek also make conductive textiles [33] that include heating products. WarmX produce heated clothing [34]. In addition to heating, a paper in 2012 described a wearable textile-based cooling system using thermoelectric modules and refrigerant channels [35].

3. Materials Developments and Wearable Computing

Wearable computing is a type of wearable that contains information technology, and is able to store, manipulate, or transmit data. The inclusion of wearable computing is what makes an E-textile a smart textile. The 1990s and early 2000s saw patents for devices either integrated onto the surface of garments or contained within pockets beginning to emerge [36–42]. These are generally regarded as the first generation E-textiles, where an electrical circuit or electronic components were attached to a garment. The initial commercialized 'first generation E-textile' garments began to appear on the market with the Industrial Clothing Division plus-jacket in 2000 [43]. This jacket attached electronic devices into pockets.

An important component towards the successful development of wearable computing included methods of creating electronic circuits within textiles. Post et al. filed a patent in the late 1990s describing how to integrate devices and circuits into textiles [44]. Another patent described using electrically insulating and electrically conductive yarns woven into a garment to create an electronic circuit [45]. A later patent further built on the use of knitting techniques to create electrical circuits and pathways [46]. In 2006 Ghosh et al. described an in-depth means of forming electrical circuits within textile structures [47]. Textilma were granted a patent describing an elastic compound thread with electrical conductivity [48]. A patent two years later from Seoul National University described a system for power transmission using conductive sewing thread [49]. An E-textile patents published in 2006 discussed the inclusion of devices such as Bluetooth and the electrical connections required between components [50]. With the concept of wearable computing existing for a number of years, in 2006 Frost and Sullivan provided an excellent review of wearable computing at that time [51].

A variety of other developments linked to wearable computing appeared in the 2000s. WearIT@work [52] was a project funded by the European Commission (€14.6 million) to investigate wearable computing. A particular focus was user acceptance. The European Space Agency (ESA) iGarment project was developed to create an Integrated System for Management of Civil Protection Units. The project was targeted to make a full-body smart garment. This would have incorporated sensors for monitoring vital signs and position. The garment was also to have included a communications unit and GPS [53].

Some recent developments in wearable computing have veered away from textiles including the Google Glass [54]. Despite these devices being wearables and not textiles, textile computing has still made significant strides in recent years.

Graphene technology may be a key contributor to the further development of wearable computing. The US Army Research Laboratory discussed the potential of graphene-based nano-electronics for applications in wearable electronics in a report produced in 2012 [55]. Interfacing with E-textiles is also of importance for future developments: A recent Microsoft patent described an interface system for using muscle movements to control a computer or other device [56].

The use of organic conductive polymers may also improve the potential for wearable computing. Hamedi et al. have developed a fiber-level electrochemical transistor, opening up the potential to create larger circuits using a weave of these fibers [57].

4. Sensors

The development of E-textiles in an academic setting first gained traction in the late 1990s with a series of publications from the Massachusetts Institute of Technology (MIT) and the Georgia Institute of Technology [58]. The 'The Wearable Motherboard^{TM}' proposed a smart-shirt capable of un-obstructively monitoring human life signs [59].

Work by Farringdon et al. described a functional fabric in the form of a woven stretch sensors later 1990s. They produced a fabric stretch sensors for the monitoring of body movement [60].

A master's thesis from the US Naval Postgraduate School in 2006 identified a further potential use of wearable sensors, suggesting that they may prove beneficial to locate sniper fire in combat situations [61].

Health and wellbeing applications also gained interest in the 2000s, with Cooseman et al. [62] describing a garment with an embedded patient monitoring system, which included wireless communication and inductive powering. In this instance, the garment was designed for infants. A paper in 2008 described a process for transforming cotton thread for use in E-textiles by applying carbon nanotubes [63]. It was claimed that the technology could be used for bio-sensing with a key result of the paper being that the carbon nano-tube cotton threads could be used to detect albumin, which is an important protein in the blood.

Work has investigated the use of textile electrodes for heart rate monitoring or ECG (Electrocardiogram). MyHeart was an EU (FP 6) funded project to develop smart fabrics for both ECG and respiration [64], with the project concluding in 2009. Another EU FP 6 project, CONTEXT, also investigated the use of textile electrodes, in this case to measure muscle and heart electric signals [65].

A paper from 2009 described an integrated temperature sensor [66]. Also in 2009, a report described the development of a knitted biomedical sensor for the monitoring body temperature [67]. A patent was granted for a linear electronic transducer for strain measurement in 2011 [68].

A proliferation of sensor embedded garments began to emerge on the market with the company Clothing+ producing both the Sensor Belt [69] and the Pure Lime sensor bra. Both devices focused on the active-wear market. At the time of writing it is unclear whether the Pure Lime sensor bra is still in production. Other active-ware devices have included the Adidas MiCoach heartrate monitoring bra. Despite interest by a number of media outlets [70], at the time of writing it appears that this product is no longer sold by Adidas.

The incorporation of sensors into garments has also continued to generate interest in the literature and from companies. A 2011 paper from described the development of a socks with integrated strain sensors for monitoring foot movement. Such systems could have had applications in stroke rehabilitation [71]. There have also been other example of incorporating temperature-sensing elements into textiles [72–75].

A variety of textile-based sensors and sensing garments have also appeared on the market. Nypro (formerly Clothing+) produces textile-based sensor systems [76]. Ohmatex [77] has created textile conductors and sensors. Polar [78] sell wearable monitoring equipment, including heartrate monitors. SmartLife [79] produce knitted physiological measurement devices for healthcare, sports and military applications. The Zephyr BioHarness [80] takes physiological meassurements from the wearer. The recorded data can then be transmitted. Under Armour have produced a shirt that monitors biometric data [81].

5. E-Textile Pressure Sensors and Textile Switches

Many early functional fabrics took the form of textile switches [82] and this area of research was expanded upon heavily by academia. The work of Post et al. included the development of a textile-based keyboard using embroidered electrodes with a silk and twisted gold spacer fabric [83]. This technology was the first step forwards pressure sensitive fabrics.

Further developments into textile transducers came in a paper from 2004 which described knitted transducers for motion and gesture capture together with ECG (electrocardiogram) measurement [84], which was also patented [85]. A related paper from 2005 described knitted capacitive transducers for touch and proximity sensing [86].

Sergio et al. initially proposed a textile based capacitive pressure sensor where a three-layer structure was implemented with a layer of rows of conductive fibers, an elastic spacer foam, and then another layer of conductive fibers orientated perpendicular to the top layer [87]. Their work described four methods for producing the conductive tracks; weaving in a mix of conductive and insulating fibers, embroidering the circuit using conductive thread, textile electrodes separated by conducting strips (called bundle routing in the paper), and the use of conductive paint. A follow-up paper described a measurement system in combination with their E-textile that was capable of producing pressure images [88]. Other researchers also developed this type of technology. Mannsfeld et al. developed a highly flexible and inexpensive capacitive pressure sensor using a micro-structured thin film as the capacitors dielectric layer [89]. Takamatsu et al. fabricated a pressure sensitive textile using a perfluoropolymer spacer, and rows of woven, die-coated yarns to make conductive rows [90]. Meyer et al. have contributed to the development of textile pressure sensors. Their general pressure sensor design consisted of three parts, an embroidered set of 2 cm × 2 cm electrodes, a 3D-knitted spacer fabric, and a woven back electrode [91,92]. Hoffmann et al. used a similar principle for a system to measure respiratory rate, where two conductive fabrics were placed on either side of a 3D-spacer textile [93]. A different proposal to monitor respiratory rate has used the change in induction of two knitted coils as the coils moved [94].

Holleczek et al. created a sensor using a pair of textile electrodes and a proprietary resin spacer material, integrating the sensor into socks [95]. Other capacitive textile designs have included a capacitive fiber by Gu et al. [96]. The capacitive fiber consisted of a conducting copper wire (0.12 mm diameter) embedded within a fiber. A similar design was employed by Lee et al. where a conductive coating was applied to a Kevlar fiber, which was then coated in a Polydimethylsiloxane (dielectric) layer. Capacitive junctions seen where the two fibers intersected [97]. Other copper wires were wrapped around the fibers surface acting as the second electrode.

Pressure sensing textiles have not been limited to a research environment. Commercially, Novel sell a number of pressure sensing products including insoles, seating covers, and gloves [98]. The XSENSOR® Technology Corporation have focused on seating sensors (in particular for the automotive industry), health care monitoring, and sleep solutions using pressure sensitive mats to aid in mattresses selection [99]. Their technology has been implemented in a variety of medical studies, with a particular focus of managing pressure ulcers [100–103]. Pressure Profile Systems continue to sell capacitive pressure sensors as of mid-2016 for a variety of applications, including gloves, robotics, pressure mats, and medical applications [104]. LG have recently announced the release of a flexible textile sensor based on capacitive technology [105]. It is difficult to ascertain the exact make-up of many of these sensors and they may not be true E-textiles. Certainly, Peterson et al. stated that the XSENSOR® devices use proprietary capacitive technology, and described the sensor as a flexible thin pad [100].

The simpler textile switch also found its way into commercial products. In 2006 an intelligent push-button system was described in a patent [106]. Beyond this a patent in 2006 [107], followed by an article [108] described a fully integrated textile switch. Other patents by France Telecom (Now Orange S.A., Paris, France)) [109], Daimler Chrysler (Auburn Hills, MI, USA) [110] and Sentrix (New York, NY, USA) [111] also describe alternative textile switch technologies.

The Burton Amp snowboarding jacket saw integrated textile switches on the arm of the jacket used to control an Apple iPod in 2002 [112]. Nike + iPod Sports Kit was another example of Apple engaging with sports apparel companies to create an E-textile device. In this case a sensor was incorporated into the shoe which communicated with an Apple product (such as an iPhone) or other Nike wearables (such as the Nike + Sportband) to track activity [113]. The recent Google and Levi's Jacquard jacket also included textile switches [114].

6. Textile Energy Solutions

By the early 2000s the first patent for textile-based energy harvesting appeared in the form of a mechanical generator scavenging energy through motion [115]. A paper by Qin et al. in 2007 described a technique for energy scavenging using piezoelectric zinc oxide nanowires grown radially around textile fibers [116]; this being one of the earlier examples of energy scavenging within textiles.

The further development of energy storage and scavenging within textiles became a key area of interest in the 2010s as the viability of many of E-textile products and concepts are dependent on a suitable power supply. Existing textile energy scavenging (or energy harvesting) devices exploit either thermoelectrics, kinetics, or photovoltaics. Each system possesses different advantages or disadvantages, with many modern systems incorporated into textiles providing flexibility but lacking other textile properties such as bend or shear. In all cases the harvested power was minimal, with the most promising generators capable of producing sustained power on the order of milliwatts. Both thermoelectric and kinetic systems draw energy from the wearer while photovoltaic energy harvesting draws energy from light sources, such as the sun. Photovoltaics have also shown significant promise concerning the power that can be generated (~30 mW/m^2) [117].

The use of carbon-nano tube based systems have gained significant popularity for both energy harvesting and storage [117–121]. For an energy storage device the textile can be used as a substrate for flexible films, or a carbon nano-tube infused film can be used to produce yarns [120]. Storage technology is either capacitive (normally supercapacitors) or chemical in nature. The use of carbon nano-tubes has caused some worry due to safety concerns [122], which would possibly impede future commercialization.

Another popular storage method has seen flexible, solid electrolyte-based batteries [123] woven into a garment as thin strips, however at present these batteries are still large relative to a normal yarn (width = 10 mm).

Triboelectric nano-generators are also viewed as a potential source of energy for wearables given their very small size, high peak power densities, and good energy conversion efficiencies [124]. Triboelectric generators convert mechanical energy and therefore could be powered by human motion or vibration. Despite high energy densities, the converted energy has a small current and the generators energy output over time is unpredictable, requiring complex supporting power management electronics. Cui et al. have demonstrated a cloth-based triboelectric generator where frictional forces between the forearm of the wearer and their body was used to generate energy [125].

Comprehensive reviews of energy harvesting and storage in textiles are available elsewhere [14,120].

7. Communication Devices

A textile-based antenna fabricated from polymer (polypyrrole) strips has been described in the literature [126]. Two conference papers from 2010 have also describe conformable antennas for space suits [127,128]. The production of flexible embroidered antennas has been reported as suitable for megahertz frequency communications [129,130]. This type of antenna is highly sensitive and has been employed for unilateral magnetic resonance measurements [131].

Incorporating radio frequency identification (commonly known as RFID) tags into textiles has also been investigated by a number of entities. Textilma were granted a patent describing an RFID module textile tag [132]. Another patent described a method of attaching RFID chips to a textile substrate [133] however, full integration within yarns was not proposed. A patent from 2007 also described a RFID device, this time focusing on clothing [134]. In 2005 a patent described the incorporation of RFID devices within epoxy resin for use in laundries [135]. A patent from 2010 described an RFID tag with integrated antenna [136] whilst another patent from the same year described a method of incorporating a RFID device into a textile tag [137].

8. Illumination

Interest in illuminated textiles continued into the mid-2000s with a patent from Daimler Chrysler in 2005 describing a textile-based lighting system for automotive applications [138]. A patent described creating a flat-panel video display by weaving electronically conductive fibers, such as dielectrics [139]. This led to patents being filed for other textile-based flexible displays in later years [140,141]. Other patents granted described a lighting system that used light leakage from optical fibers in specific locations [142] to create illumination.

A patent from 2011 described a method of producing an illuminated pattern using light conducting fibres [143]. Another illuminated textile using optical fibres was described in a patent from the same year [144]. The company Sensing Tex Sl have used optical fibers to illuminate textiles [145].

In contrast, Philips have produced illuminated textiles using LED technology [146]. Philips have previously patented a flexible electro-optic filament [147]. There are a variety of companies that produce illuminated clothing for fashion applications, as well as use in performance (i.e. theatre), these include Cutecircuit [148] and LUcentury [149] which create garments by sewing LEDs onto existing clothing. Given advances in LED technology, reducing their size, LEDs can easily be incorporated into yarns using functional electronic yarns technology, and subsequently into garments [75]. Electroluminescent yarn was also developed on 2010 [150] with the technology fully described in a 2012 paper [151]. Another form of lighting used for E-textiles was by attaching lasers to a garment, as shown by Bono in 2009 [152]. This technology is not practical for a mass produced garment for a variety of reasons including cost and the weight of the final garment.

9. Market Trends

According to IMS Research (Wellingborough, UK), 14 million wearable devices were shipped in 2011 [153]. This clear market coupled with the significant scientific advances in E-textiles led to a proliferation of devices entering the market in the early 2010's, and an increased interest in the technology from academia. More recently the Fung Global Retail and Technology report on wearables in 2016 [154] identified a significant increase in the wearable market between 2015 and 2016, a 18.4% climb to $28.7 billion. This is far higher than the predictions of the IMS, expecting that the revenue for wearable technology would be $6 billion in 2016. Forbes currently predict that the wearable market will reach $34 billion by 2020 [155]. It is of interest to note that the Fung report also states that the top wearable brands of 2016 were Fitbit, Xiaomi, and Apple based on the number of units shipped: These company's products are not textile based. The principal wearable device sold by each company are smart watches. It is unclear whether the growing market for wearables will become more focused on textile-based devices in the coming years, but the advantages offered, such as comfort to the wearer, will likely be an influential factor. The current market for E-textiles generates sales of around $100 million per year, with some sources predicting a $5 billion market by 2027 [156].

Current commercially available E-textiles include soft textile switches produced by International Fashion Machines (Seattle, WA, USA) [157] and systems for incorporating wires into clothing by the Technology Enabled Clothing [158]. Additionally, Fibretronic [159] have created a varied range of products including (but not limited to) flexible switches, textile cables for signal or power transport, and textile sensors, however as of 2018 it is unclear whether they are still trading. As discussed earlier some sportswear E-textile products are no longer available, presumably due to poor market demand. An important thing to note is that a significant number of website sources have been using in this review, particularly regarding commercial products, as information was not available from other sources. This is another possible indicator of poor uptake of certain products.

10. Potential Future Developments

While this review is focused on the history and evolution of E-textiles, it is of interest to consider the direction that E-textile research will take in upcoming years. Originally, the vision of those working in the field of E-textiles was to incorporate all of the required electronic systems within the textile. More recently however some claim that the best approach is to use mobile phones as an interface [160]; which has been aided in part by the substantial advancement in mobile phone technology in recent years. Others claim that this approach is a temporary diversion and there are many advantages for fully embedded systems, and that as developments progress, the mobile phone itself will be integrated into textiles. A complete integration of the electronics may also be unfavorable for sustainability reasons as it makes the electronics more difficult to remove at the end of the life of a product [161].

It is expected that there will be a far greater uptake of wearable electronics when battery technology is improved or alternative energy sources, such as energy scavenging, become more viable. This is of particular importance to E-textiles over wearables more generally, as most conventional power sources are not well suited to textile integration due to size, inflexibility, and lack of washability. The reductions in size and cost of components will promote further development and uptake. A review paper from 2012 discussed the subject [162] and a BBC news item outlined on-going UK research in the area [163].

Ultimately, the adoption of E-textiles will depend on the cost. This will reduce with material costs and improvements to the manufacturing processes. It is also possible that developments in graphene technology will be improve the potential of what can be achieved with electronic textiles. Many major companies including Samsung, Nokia, and IBM have made significant investments into graphene technology [164]. There is the potential that graphene's physical properties, including its strength and electrical conductivity, will allow it to replace silicon in many devices, possibly by the late 2020s after the technology has matured. In addition, work on carbon nanotubes is beginning to show some promise, especially for energy-based applications (such as energy harvesting and scavenging) [117,119–121]. Both technologies offer potential for further miniaturization of embedded electronics.

With an enhancement of how much can be fit within a textile, and suitable energy solutions, E-textiles could move towards true wearable computing, with the textile managing and processing data on its own depending upon requirements. The decreasing size of microprocessors makes embedding this kind of intelligence within a textile likely in the immediate future.

11. Conclusions

This review of the literature has clearly shown that the three pathways of integrating electronics into textiles have been applied in different ways. The methods of integrating electronics offer different advantages and disadvantages. The first generation E-textiles will always interfere with the textile properties of a garment, even thin film devices (while flexible), will not possess the shear properties of a normal textile. The second generation textiles may retain a textile feel but are limited in their applications; such as the creation of electronic pathways, and electrode-based sensing.

The third generation of E-textiles, where electronics are contained within the yarn structure, do not interfere with the textile properties of a fabric. As this technology is principally limited by the size of the incorporated electronics (i.e., the electronic chip dimensions) the potential of this area will grow as smaller electronic chips become available.

While the history of E-textiles has shown the development of new techniques to integrate electronics within a textile it is likely that the existing three methods will remain in use into the future. The attachment of electronics onto a garment is still common, in particular for illuminated textiles, despite this technology first being demonstrated in 1883.

Author Contributions: T.H.-R., C.C. and T.D. located the reference material used in the review. T.H.-R. and T.D. reviewed all of the reference material used in this review. T.H.-R. prepared the final manuscript.

Conflicts of Interest: The authors declare no conflict of interest.

References

1. Adovasio, J.M.; Soffer, O.; Klíma, B. Upper Palaeolithic fibre technology: Interlaced woven finds from Pavlov I, Czech Republic, c. 26,000 years ago. *Antiquity* **1996**, *70*, 526–534. [CrossRef]
2. Lewis, P. William Lee's stocking frame: Technical evolution and economic viability 1589–1750. *Text. Hist.* **1986**, *17*, 129–147. [CrossRef]
3. Thackeray, F.W.; Findling, J.E. (Eds.) *Events that Changed Great Britain Since 1689*; Greenwood Publishing Group: Westport, CT, USA, 2002.
4. Guler, S.D.; Gannon, M.; Sicchio, K. A Brief History of Wearables. In *Crafting Wearables*; Apress: New York, NY, USA, 2016; pp. 3–10.
5. Shirakawa, H.; Louis, E.J.; MacDiarmid, A.G.; Chiang, C.K.; Heeger, A.J. Synthesis of electrically conducting organic polymers: Halogen derivatives of polyacetylene,$(CH)_x$. *J. Chem. Soc. Chem. Commun.* **1977**, *16*, 578–580. [CrossRef]
6. The Nobel Prize in Chemistry 2000. Available online: https://www.nobelprize.org/nobel_prizes/chemistry/laureates/2000/ (accessed on 21 August 2017).
7. Paton, G.A.; Sterling, M.N.; Sanders, J.H. Integral, Electrically-Conductive Textile Filament. U.S. Patent 4,045,949, 6 September 1977.
8. 1960: Metal Oxide Semiconductor (MOS) Transistor Demonstrated, The Silicon Engine, Computer History Museum. Available online: http://www.computerhistory.org/siliconengine/metal-oxide-semiconductor-mos-transistor-demonstrated/ (accessed on 21 August 2017).

9. Miller, G.E.; Dalke, M. Illuminated Article of Clothing. U.S. Patent 4,164,008, 7 August 1979.
10. Dias, T.; Fernando, A. Operative Devices Installed in Yarns. GB0509963B1, 9 May 2005.
11. Dias, T. Electronic strip yarn. WO2017/1150873 A1, 6 July 2017.
12. Dias, T.K.; Rathnayake, A. Electronically Functional Yarns. GB2529900, 9 March 2016.
13. Conroy, D.W.; García, A. A golden garment from ancient Cyprus? Identifying new ways of looking at the past through a preliminary report of textile fragments from the Pafos 'Erotes' Sarcophagus. In *The SInet 2010 eBook*; University of Wollongong: Wollongong, Australia, 2010; p. 36.
14. Stoppa, M.; Chiolerio, A. Wearable electronics and smart textiles: A critical review. *Sensors* **2014**, *14*, 11957–11992. [CrossRef] [PubMed]
15. Zeng, W.; Shu, L.; Li, Q.; Chen, S.; Wang, F.; Tao, X.M. Fiber-based wearable electronics: A review of materials, fabrication, devices, and applications. *Adv. Mater.* **2014**, *26*, 5310–5336. [CrossRef] [PubMed]
16. Weng, W.; Chen, P.; He, S.; Sun, X.; Peng, H. Smart electronic textiles. *Angew. Chem. Int. Ed.* **2016**, *55*, 6140–6169. [CrossRef] [PubMed]
17. Tao, X. (Ed.) *Handbook of Smart Textiles*; Springer: Singapore, 2015; ISBN 9789814451468.
18. Dias, T. (Ed.) *Electronic Textiles: Smart Fabrics and Wearable Technology*; Woodhead Publishing: Cambridge, UK, 2015; ISBN 9780081002018.
19. Carron, A.L. Electric-Heated Glove. U.S. Patent 1,011,574, 9 September 1911.
20. Joule, J.P. On the Production of Heat by Voltaic Electricity. *Proc. R. Soc.* **1840**, *1*, 59–60. [CrossRef]
21. Pollak, A. Electrically-Heated Garment. U.S. Patent 1,073,926, 23 September 1913.
22. Lemercier, A.A. Electrically-Heated Clothing. U.S. Patent 1,284,378, 12 November 1918.
23. Heinemann, O. Electrically-Heated Boot. U.S. Patent 1,761,829, 3 June 1930.
24. Benjamin, B. Electrically-Heated Garment. U.S. Patent 1,358,509, 9 November 1920.
25. Graham, W.D.; Uhlig, C.M. Electrically-Heated Garment. U.S. Patent 1,691,472, 13 November 1928.
26. Hefter, M. Knitted Electric Heating-Body. U.S. Patent 975,359, 8 November 1910.
27. Grisley, F. Improvements in Blankets, Pads, Quilts, Clothing, Fabric, or the Like, Embodying Electrical Conductors. GB445195(A), 30 March 1936.
28. Ellis, H.G. Heated Baby Carriage Blanket. U.S. Patent 2,993,979, 25 July 1961.
29. Balz, C.F.; Murphy, D.J. Electrically Heated Sock with Battery Supporting Pouch. U.S. Patent 3,396,264, 6 August 1968.
30. Pavitt, J. *Fear and Fashion in the Cold War*; V&A Publishing: London, UK, 2008.
31. Bourke, M.J.; Clothier, B.L. Inductively Heated Clothing. WO2008101203, 21 August 2008.
32. EXO^2. Available online: http://www.exo2.co.uk/ (accessed on 22 August 2017).
33. Marktek Inc. EMI Shielding, Conductive, Resistive and Radar Absorptive Materials. Available online: http://www.marktek-inc.com/ (accessed on 22 August 2017).
34. Beheizbare Unterwäsche als Alltagslösung von. warmx.de. Available online: http://www.warmx.de/index.php (accessed on 22 August 2017).
35. Delkumburewattea, G.B.; Dias, T. Wearable cooling system to manage heat in protective clothing. *J. Text. Inst.* **2012**, *103*, 483–489. [CrossRef]
36. Carroll, D.W. Wearable Personal Computer System. U.S. Patent 5,555,490, 10 September 1996.
37. Egan, E.; Amon, C.H. Cooling strategies for embedded electronic components of wearable computers fabricated by shape deposition manufacturing. In Proceedings of the I-THERM V., Inter-Society Conference on Thermal Phenomena in Electronic Systems, Orlando, FL, USA, 29 May–1 June 1996; pp. 13–20.
38. Baudhuin, E.S. Telemaintenance applications for the Wearable/sup TM/PC. In Proceedings of the 15th AIAA/IEEE, Digital Avionics Systems Conference, Atlanta, GA, USA, 31 October 1996; pp. 407–413.
39. Janik, C.M. Flexible Wearable Computer. U.S. Patent 6,108,197, 22 August 2000.
40. Tento, H.; Nasu, R. Base for Wearable Computer. JP2000357025, 26 December 2000.
41. Dale, C.A. Wearable Computer Apparatus. U.S. Patent 6,167,413, 26 December 2000.
42. Jenkins, M.D. Convertible Wearable Computer. HK1024069, 22 September 2006.
43. The ICD+ jacket: Slip into My Office, Please. The Independent. 4 September 2000. Available online: http://www.independent.co.uk/news/business/analysis-and-features/the-icd-jacket-slip-into-my-office-please-694074.html (accessed on 17 November 2017).
44. Post, E.R.; Orth, M.; Cooper, E.; Smith, J.R. Electrically Active Textiles and Articles Made Therefrom. U.S. Patent 6,210,771, 3 April 2001.

45. Hill, I.G.; Trotz, S.; Riddle, G.H.N.; Brookstein, D.S.; Govindaraj, M. Plural Layer Woven Electronic Textile, Article and Method. U.S. Patent 7,144,830, 5 December 2006.
46. Dias, T.K.; Mitcham, K.; Hurley, W. Knitting Techniques. U.S. Patent 7,779,656, 24 August 2010.
47. Ghosh, T.K.; Dhawan, A.; Muth, J.F. Formation of electrical circuits in textile structures. In *Intelligent Textiles and Clothing*; North Carolina State University: Raleigh, NC, USA, 2006.
48. Speich, F. Electrically Conductive, Elastic Compound Thread, Particularly for RFID Textile Labels, the Use Thereof, and the Production of a Woven Fabric, Knitted Fabric, or Meshwork Therewith. TW200840891, 16 October 2008.
49. Kang, T.J.; Kim, B.D. Electrically Conductive Sewing Thread for Power and Data Transmission Line of Smart Interactive Textile Systems. KR20090012769, 4 February 2009.
50. Klaus, P.; Jochen, D.; Horst, T. Article of Clothing for Attaching e.g., ID Systems and Blue Tooth Modules Comprises Electrical Internal and External Components Arranged on the Surface of the Article so that Electrical Lines Extend between the External Components. DE102004039765, 9 March 2006.
51. Wearable Computing (Technical Insights). Available online: http://www.frost.com/sublib/display-report.do?id=D626-01-00-00-00 (accessed on 21 August 2017).
52. WearIT@Work Project-Customer Cases-Xsens 3D Motion Tracking. Available online: https://www.xsens.com/customer-cases/wearitwork-project/ (accessed on 21 August 2017).
53. Final Presentation of the I-GARMENT Project. Available online: http://www.esa.int/Our_Activities/Preparing_for_the_Future/Space_for_Earth/Space_for_health/Final_presentation_of_the_I-GARMENT_project (accessed on 21 August 2017).
54. Google's Project Glass Made Available to Developers. Available online: https://www.theguardian.com/technology/2012/jun/28/google-project-glass-available-to-developers (accessed on 21 August 2017).
55. Dubey, M.; Nambaru, R.; Ulrich, M.; Ervin, M.; Nichols, B.; Zakar, E.; Nayfeh, O.M.; Chin, M.; Birdwell, G.; O'Regan, T. *Graphene-Based Nanoelectronics*; U.S. Army Research Laboratory: Adelphi, MD, USA, 2012.
56. Tan, D.; Saponas, T.S.; Morris, D.; Turner, J. Wearable Electromyography-Based Human-Computer Interface. US2012188158, 26 July 2012.
57. Hamedi, M.; Forchheimer, R.; Inganäs, O. Towards woven logic from organic electronic fibres. *Nat. Mater.* **2007**, *6*, 57–362. [CrossRef] [PubMed]
58. Post, E.R.; Orth, M. Smart Fabric, or Washable Computing. In Proceedings of the First IEEE International Symposium on Wearable Computers, Cambridge, MA, USA, 13–14 October 1997.
59. Gopalsamy, C.; Park, S.; Rajamanickam, R.; Jayaraman, S. The Wearable Motherboard™: The first generation of adaptive and responsive textile structures (ARTS) for medical applications. *Virtual Real.* **1999**, *4*, 152–168. [CrossRef]
60. Farringdon, J.; Moore, A.J.; Tilbury, N.; Church, J.; Biemon, P.D. Wearable sensor badge and sensor jacket for context awareness. In Proceedings of the Third International Symposium on Wearable Computers, Digest of Papers, San Francisco, CA, USA, 18–19 October 1999; pp. 107–113.
61. Stephen, T.K.S. Source Localization Using Wireless Sensor Networks. Thesis for Master of Science in Electrical Engineering, Naval Postgraduate School Monterey, Monterey, CA, USA, 2006. Available online: https://calhoun.nps.edu/bitstream/handle/10945/2689/06_Jun_Tan.pdf?sequence=1 (accessed on 15 May 2018).
62. Coosemans, J.; Hermans, B.; Puers, R. Integrating wireless ECG monitoring in textiles. *Sens. Actuators A Phys.* **2006**, *130*, 48–53. [CrossRef]
63. Shim, B.S.; Chen, W.; Doty, C.; Xu, C.; Kotov, N.A. Smart Electronic Yarns and Wearable Fabrics for Human Biomonitoring made by Carbon Nanotube Coating with Polyelectrolytes. *Nano Lett.* **2008**, *8*, 4151–4157. [CrossRef] [PubMed]
64. European Commission: CORDIS: Projects and Results: MyHeart. Available online: https://www.cordis.europa.eu/project/rcn/71193_en.html (accessed on 25 April 2018).
65. European Commission: CORDIS: Projects and Results: Contactless Sensors for Body Monitoring Incorporated in Textiles. Available online: https://cordis.europa.eu/project/rcn/80730_en.html (accessed on 25 Apri 2018).
66. Kinkeldei, T.; Zysset, C.; Cherenack, K.; Troester, G. Development and evaluation of temperature sensors for textile integration. In Proceedings of the IEEE Sensors, Christchurch, New Zealand, 25–28 October 2009.
67. Husain, M.D.; Dias, T. Development of Knitted Temperature Sensor (KTS). Available online: http://www.systex.eu/sites/default/files/Systex_award_2009_Dawood_03Sept09.pdf (accessed on 15 May 2018).

68. Challis, S.; Fernando, A.; Dias, T.; Cooke, W.; Chaudhury, N.H.; Geraghty, J.; Smith, S. Precision Delivery System. US2003110812, 19 June 2003.
69. Reho, A. Clothing+. In Proceedings of the Smart Fabrics Conference, Miami, FL, USA, 17-19 April 2012.
70. Best Heart Rate Monitor Sports Bra that's actually Smart and Comfortable. Available online: http://notsealed.com/heart-rate-monitor-sports-bra-smart-comfortable.html (accessed on 17 November 2017).
71. Preece, S.J.; Kenney, L.P.J.; Major, M.J.; Dias, T.; Lay, E.; Fernandes, B.T. Automatic identification of gait events using an instrumented sock. *J. NeuroEng. Rehabil.* **2011**, *8*, 32. [CrossRef] [PubMed]
72. Hughes-Riley, T.; Lugoda, P.; Dias, T.; Trabi, C.L.; Morris, R. A study of thermistor performance within a textile structure. *Sensors* **2017**, *17*, 1804. [CrossRef] [PubMed]
73. Cherenack, K.; Zysset, C.; Kinkeldei, T.; Münzenrieder, N.; Tröster, G. Woven electronic fibers with sensing and display functions for smart textiles. *Adv. Mater.* **2010**, *22*, 5178–5182. [CrossRef] [PubMed]
74. Lugoda, P.; Dias, T.; Morris, R. Electronic temperature sensing yarn. *J. Multidiscip. Eng. Sci. Stud.* **2015**, *1*, 100–103.
75. Dias, T.; Hughes-Riley, T. Electronically Functional Yarns Transform Wearable Device Industry. *Read. Res. Dev. Commun.* **2017**, *59*, 19–21.
76. Consumer Health. Available online: https://www.jabil.com/solutions/by-industry/healthcare/consumer-health.html (accessed on 22 August 2017).
77. Ohmatex. Available online: http://www.ohmatex.dk/ (accessed on 22 August 2017).
78. Heart Rate Monitors, Activity Trackers and Bike Computers. Available online: https://www.polar.com/en (accessed on 22 August 2017).
79. Smartlife. Available online: https://www.smartlifeinc.com/ (accessed on 22 August 2017).
80. Zephyr™ Performance Systems. Performance Monitoring Technology. Available online: https://www.zephyranywhere.com/ (accessed on 22 August 2017).
81. Under Armour-Sportswear, Sport Shoes, & Accessories. Available online: http://www.underarmour.co.uk/ (accessed on 22 August 2017).
82. Schwabe, D. Electrical Switch for Automotive or Clothing Use is Integrated into a Flexible Textile Material and is Actuated by Applied Force. DE102005038988 (A1), 22 February 2007.
83. Post, E.R.; Orth, M.; Russo, P.R.; Gershenfeld, N. E-broidery: Design and fabrication of textile-based computing. *IBM Syst. J.* **2000**, *39*, 840–860. [CrossRef]
84. Wijesiriwardana, R.; Mitcham, K.; Dias, T. Fibre-meshed transducers based real time wearable physiological information monitoring system. In Proceedings of the ISWC 2004. Eighth International Symposium on Wearable Computers, Arlington, VA, USA, 31 October–3 November 2004; Volume 1, pp. 40–47.
85. Tilak, D.; Beatty, P.C.W.; Cooke, W.; Wijesiriwardana, R.; Mitcham, K.; Mukhopadhyay, S.; Hurley, W. Knitted Transducer Devices. U.S. Patent Application 10/557,074, 19 May 2004.
86. Wijesiriwardana, R.; Mitcham, K.; Hurley, W.; Dias, T. Capacitive fiber-meshed transducers for touch and proximity-sensing applications. *IEEE Sens. J.* **2005**, *5*, 989–994. [CrossRef]
87. Sergio, M.; Manaresi, N.; Tartagni, M.; Guerrieri, R.; Canegallo, R. A textile based capacitive pressure sensor. In Proceedings of the IEEE Sensors, Orlando, FL, USA, 12–14 June 2002; Volume 2, pp. 1625–1630.
88. Sergio, M.; Manaresi, N.; Campi, F.; Canegallo, R.; Tartagni, M.; Guerrieri, R. A dynamically reconfigurable monolithic CMOS pressure sensor for smart fabric. *IEEE J. Solid-State Circuits* **2003**, *38*, 966–975. [CrossRef]
89. Mannsfeld, S.C.; Tee, B.C.; Stoltenberg, R.M.; Chen, C.V.H.; Barman, S.; Muir, B.V.; Sokolov, A.N.; Reese, C.; Bao, Z. Highly sensitive flexible pressure sensors with microstructured rubber dielectric layers. *Nat. Mater.* **2010**, *9*, 859–864. [CrossRef] [PubMed]
90. Takamatsu, S.; Kobayashi, T.; Shibayama, N.; Miyake, K.; Itoh, T. Fabric pressure sensor array fabricated with die-coating and weaving techniques. *Sens. Actuators A Phys.* **2012**, *184*, 57–63. [CrossRef]
91. Meyer, J.; Lukowicz, P.; Troster, G. Textile pressure sensor for muscle activity and motion detection. In Proceedings of the 10th IEEE International Symposium on Wearable Computers, Montreux, Switzerland, 11–14 October 2006; pp. 69–72.
92. Meyer, J.; Arnrich, B.; Schumm, J.; Troster, G. Design and modeling of a textile pressure sensor for sitting posture classification. *IEEE Sens. J.* **2010**, *10*, 1391–1398. [CrossRef]
93. Hoffmann, T.; Eilebrecht, B.; Leonhardt, S. Respiratory monitoring system on the basis of capacitive textile force sensors. *IEEE Sens. J.* **2011**, *11*, 1112–1119. [CrossRef]

94. Wijesiriwardana, R. Inductive fiber-meshed strain and displacement transducers for respiratory measuring systems and motion capturing systems. *IEEE Sens. J.* **2006**, *6*, 571–579. [CrossRef]
95. Holleczek, T.; Rüegg, A.; Harms, H.; Tröster, G. Textile pressure sensors for sports applications. In Proceedings of the IEEE Sensors, Kona, HI, USA, 1–4 November 2010; pp. 732–737.
96. Gu, J.F.; Gorgutsa, S.; Skorobogatiy, M. A Fully Woven Touchpad Sensor Based on Soft Capacitor Fibers. Unpublished. 2011. Available online: https://www.researchgate.net/profile/Maksim_Skorobogatiy/publication/51912313_A_fully_woven_touchpad_sensor_based_on_soft_capacitor_fibers/links/00b4951cad5dfa7912000000.pdf (accessed on 4 May 2016).
97. Lee, J.; Kwon, H.; Seo, J.; Shin, S.; Koo, J.H.; Pang, C.; Son, S.; Kim, J.H.; Jang, Y.H.; Kim, D.E.; et al. Conductive Fiber-Based Ultrasensitive Textile Pressure Sensor for Wearable Electronics. *Adv. Mater.* **2015**, *27*, 2433–2439. [CrossRef] [PubMed]
98. Sensors. Available online: http://www.novel.de/novelcontent/sensors (accessed on 21 August 2017).
99. Xsensor Technology. Available online: http://www.xsensor.com (accessed on 21 August 2017).
100. Peterson, M.J.; Gravenstein, N.; Schwab, W.K.; van Oostrom, J.H.; Caruso, L.J. Patient repositioning and pressure ulcer risk-monitoring interface pressures of at-risk patients. *J. Rehabil. Res. Dev.* **2013**, *50*, 477. [CrossRef] [PubMed]
101. Kahn, J.A.; Kerrigan, M.V.; Gutmann, J.M.; Harrow, J.J. Pressure ulcer risk of patient handling sling use. *J. Rehabil. Res. Dev.* **2015**, *52*, 291.
102. Wong, H.; Kaufman, J.; Baylis, B.; Conly, J.M.; Hogan, D.B.; Stelfox, H.T.; Southern, D.A.; Ghali, W.A.; Ho, C.H. Efficacy of a pressure-sensing mattress cover system for reducing interface pressure: Study protocol for a randomized controlled trial. *Trials* **2015**, *16*, 1. [CrossRef] [PubMed]
103. Higer, S.; James, T. Interface pressure mapping pilot study to select surfaces that effectively redistribute pediatric occipital pressure. *J. Tissue Viability* **2016**, *25*, 41–49. [CrossRef] [PubMed]
104. Capacitive Tactile Pressure Sensors. Available online: http://www.pressureprofile.com/capacitive-sensors (accessed on 21 August 2017).
105. LG Innotek Unveils Flexible Textile Pressure Sensors. Available online: http://m.phys.org/news/2016-07-lg-innotek-unveils-flexible-textile.html?utm_source=nwletter&utm_medium=email&utm_campaign=daily-nwletter (accessed on 21 August 2017).
106. Schedukat, N.; Gries, T. Intelligent Push-Button System for Use in Smart Textile, Has Upper and Lower Push-Button Halves with Two Electric Contacts Connected with One Another Electro-Conductively for Data, Signal and Power Transmission, While Closing Connection. DE102004026554, 16 March 2006.
107. Dias, T.; Hurley, W.; Wijesiriwardana, R. Switches in Textile Structures. WO2006045988, 4 May 2006.
108. Dias, T.; Hurley, W.; Monaragala, R.; Wijeyesiriwardana, R. Development of Electrically Active Textiles. In *Advances in Science and Technology*; Trans Tech Publications: Zürich, Switzerland, 2008; Volume 60.
109. Deflin, E.; Weill, A.; Bonfiglio, J.; Athimon-Pillard, B. Flexible Textile Structure for Producing Electric Switches. WO03050832, 19 June 2003.
110. Kuebler, S.; Seidel, F.-P. Textile with Built-in Electrical Switches is Used as Internal Lining or Seat Covering in Vehicles. DE102004009189, 15 September 2005.
111. Leftly, S.A. Switches and Devices for Textile Articles. WO2006030230, 23 March 2006.
112. Greenfield, A. *Readings from Everyware: The dawning age of Ubiquitous Computing*; New Rider: San Francisco, CA, USA, 2006.
113. Nike + iPod Sensor. Available online: https://manuals.info.apple.com/MANUALS/1000/MA1139/en_US/nike_plus_ipod_sensor.pdf (accessed on 8 August 2017).
114. Jacquard by Google. Available online: https://atap.google.com/jacquard/ (accessed on 17 November 2017).
115. Muglia, H.A.; Refeld, J.; Eiselt, H. Generator Device for Converting Motion Energy of Person's Respiration into Electrical Energy is Integrated into Clothing Item Normally Arranged at One or More Positions on Person that Undergoes Change in Dimensions during Respiration. DE10340873, 28 April 2005.
116. Qin, Y.; Wang, X.; Wang, Z.L. Microfibre-nanowire hybrid structure for energy scavenging. *Nature* **2008**, *451*, 809–813. [CrossRef] [PubMed]
117. Velten, J.; Kuanyshbekova, Z.; Göktepe, Ö.; Göktepe, F.; Zakhidov, A. Weavable dye sensitized solar cells exploiting carbon nanotube yarns. *Appl. Phys. Lett.* **2013**, *102*, 203902. [CrossRef]

118. Uddin, M.J.; Davies, B.; Dickens, T.J.; Okoli, O.I. Self-aligned carbon nanotubes yarns (CNY) with efficient optoelectronic interface for microyarn shaped 3D photovoltaic cells. *Solar Energy Mater. Solar Cells* **2013**, *115*, 166–171. [CrossRef]
119. Meng, Y.; Zhao, Y.; Hu, C.; Cheng, H.; Hu, Y.; Zhang, Z.; Shi, G.; Qu, L. All-graphene core-sheath microfibers for all-solid-state, stretchable fibriform supercapacitors and wearable electronic textiles. *Adv. Mater.* **2013**, *25*, 2326–2331. [CrossRef] [PubMed]
120. Jost, K.; Dion, G.; Gogotsi, Y. Textile energy storage in perspective. *J. Mater. Chem. A* **2014**, *2*, 10776–10787. [CrossRef]
121. Zhang, D.; Miao, M.; Niu, H.; Wie, Z. Core-spun carbon nanotube yarn supercapacitors for wearable electronic textiles. *Acs Nano* **2014**, *8*, 4571–4579. [CrossRef] [PubMed]
122. Greenemeier, L. Study says carbon nanotubes as dangerous as asbestos. *Sci. Am.* **2008**, *20*. Available online: https://www.scientificamerican.com/article/carbon-nanotube-danger/ (accessed on 8 August 2017).
123. Liu, Y.; Gorgutsa, S.; Santato, C.; Skorobogatiy, M. Flexible, solid electrolyte-based lithium battery composed of LiFePO$_4$ cathode and Li$_4$Ti$_5$O$_{12}$ anode for applications in smart textiles. *J. Electrochem. Soc.* **2012**, *159*, A349–A356. [CrossRef]
124. Fan, F.R.; Tian, Z.Q.; Wang, Z.L. Flexible triboelectric generator. *Nano Energy* **2012**, *1*, 328–334. [CrossRef]
125. Cui, N.; Liu, J.; Gu, L.; Bai, S.; Chen, X.; Qin, Y. Wearable triboelectric generator for powering the portable electronic devices. *ACS Appl. Mater. Interfaces* **2015**, *7*, 18225–18230. [CrossRef] [PubMed]
126. Pillai, P.; Paster, E.; Montemayor, L.; Benson, C.; Hunter, I.W. *Development of Soldier Conformable Antennae Using Conducting Polymers*; Massachusetts Inst of Tech Cambridge Institute for Soldier Nanotechnologies (ISN): Cambridge, MA, USA, 2010.
127. Campbell, T.G.; Hearn, C.W.; Reddy, C.J.; Boyd, R.C.; Yang, T.; Davis, W.A.; Persans, A.; Scarborough, S. Development of Conformal Space Suit Antennas for Enhanced EVA Communications and Wearable Computer Applications. In Proceedings of the 2010 Antenna Applications Symposium Volume II of II, Tangshan, China, 15–18 October 2010.
128. Yang, T.; Davis, W.A.; Campbell, T.G.; Reddy, C.J. A Low-Profile Antenna Design Approach for Conformal Space Suit and Other Wearable Applications. In Proceedings of the 2010 Antenna Applications Symposium Volume II of II, Monticello, IL, USA, 21–23 September 2010.
129. Acti, T.; Zhang, S.; Chauraya, A.; Whittow, W.; Seager, R.; Dias, T.; Vardaxoglou, Y. High performance flexible fabric electronics for megahertz frequency communications. In Proceedings of the Antennas and Propagation Conference (LAPC), 2011 Loughborough, Loughborough, UK, 14–15 November 2011; pp. 1–4.
130. Chauraya, A.; Zhang, S.; Whittow, W.; Acti, T.; Seager, R.; Dias, T.; Vardaxoglou, Y.C. Addressing the challenges of fabricating microwave antennas using conductive threads. In Proceedings of the 6th European Conference on Antennas and Propagation (EUCAP), Prague, Czech Republic, 26–30 March 2012; pp. 1365–1367.
131. Morris, R.H.; McHale, G.; Dias, T.; Newton, M.I. Embroidered coils for magnetic resonance sensors. *Electronics* **2013**, *2*, 168–177. [CrossRef]
132. Speich, F. RFID Transponder Chip Module with Connecting Means for an Antenna, Textile Tag with an RFID Transponder Chip Module, and Use of an RFID Transponder Chip Module. TW200905574, 1 February 2009.
133. Muehlbauer, A.G. Method for Attaching and Contacting RFID Chip Modules to Produce Transponders Comprising a Textile Substrate, and Transponder for Fabrics. WO2007104634, 20 September 2007.
134. Corbett, B.G. Textile Identification System with RFID Tracking. US2005183990, 25 August 2005.
135. Gravina, D. Method Using RFID Technology for Surveillance of Textile Goods in Laundries. EP1528504, 4 May 2005.
136. Shpajkh, F. Textile RFID Label. RU2009114415, 27 October 2010.
137. Speich, F. Method for the Production of a Textile Label Having an RFID Transponder Chip and Interlaced Information Carrier, and System for Carrying out the Method. US2010085166, 8 April 2010.
138. Boll, W. Illumination System for Automobile Passenger Compartment e.g., for Cabriolet Automobile, Using Flexible Light Conductors or Electrical Lighting Devices Incorporated in Textile Material Forming Automobile Roof. DE10345002, 21 April 2005.
139. Christensen, A.O. Woven Polymer Fiber Video Displays with Improved Efficiency and Economy of Manufacture. U.S. Patent US 6,229,259, 8 May 2001.
140. Murasko, M.; Kinlen, P.J. Illuminated Display System and Process. U.S. Patent US 6,811,895, 2 November 2004.

141. De-Flin, E.; Mourot, E.; Remy, M. Textile Display. WO 2004/100111 A2, 18 November 2004.
142. DO UK HO. Self-Lighting Textile Using Optical Fiber. KR20080040815, 9 May 2008.
143. Peng, C.-T.; Wang, C.-T. Textile with Pattern-Lighting Effect. US2011309768, 22 December 2011.
144. Yu, Z. Lighting Textile Fabric. CN201873891, 22 June 2011.
145. Ridao, M. Self Illuminating Spaces. In Proceedings of the Smart Fabrics Conference, Miami, FL, USA, 17–19 April 2012.
146. Van De Pas, L. Bring Spaces Alive. In Proceedings of the Smart Fabrics Conference, Miami, FL, USA, 17–19 April 2012.
147. Eves, D.A.; Chapman, J.A.; Bechtel, H.-H.; Wagner, P.C.; Martynov, Y. Electro-Optic Filament or Fibre. WO/2004/055576, 1 July 2004.
148. Cutecircuit. Available online: http://cutecircuit.com/ (accessed on 22 August 2017).
149. Lucentury. Available online: http://www.lucentury.com/ (accessed on 22 August 2017).
150. Dias, T.; Monaragala, R.M. Electro-luminant Fabric Structures. US2010003496, 7 January 2010.
151. Dias, T.; Monaragala, R. Development and analysis of novel electroluminescent yarns and fabrics for localized automotive interior illumination. *Text. Res. J.* **2012**, *82*, 1164–1176. [CrossRef]
152. Bono's Laser Stage Suit by Moritz Waldemeyer. Available online: https://www.dezeen.com/2010/02/28/bonos-laser-stage-suit-by-moritz-waldemeyer/ (accessed on 14 November 2017).
153. *World Market for Wearable Technology—A Quantitative Market Assessment—2012*; IMS Research of Wellingborough: Wellingborough, UK, 2012.
154. The Wearables Report 2016: Reviewing a Fast-Changing Market. Available online: https://www.fbicgroup.com/sites/default/files/The%20Wearables%20Report%202016%20by%20FBIC%20Global%20Retail%20and%20Technology%20June%2021%202016.pdf (accessed on 22 August 2017).
155. Wearable Tech Market to be Worth $34 Billion By 2020. Available online: https://www.forbes.com/sites/paullamkin/2016/02/17/wearable-tech-market-to-be-worth-34-billion-by-2020/#5f690ca83cb5 (accessed on 22 August 2017).
156. Hayward, J. E-Textiles 2017–2027: Technologies, Markets, Players. Available online: https://www.idtechex.com/research/reports/e-textiles-2017-2027-technologies-markets-players-000522.asp (accessed on 17 November 2017).
157. International Fashion Machines. Available online: http://www.ifmachines.com/ (accessed on 21 August 2017).
158. ScotteVest. Available online: https://www.scottevest.com/ (accessed on 21 August 2017).
159. Fibretronic. Wearable Tech. CrunchWear. Available online: http://crunchwear.com/category/companies/fibretronic/ (accessed on 21 August 2017).
160. Russell, B. Smart Fabrics for Consumer Health. In Proceedings of the Smart Fabrics Conference, Miami, FL, USA, 17–19 April 2012.
161. Köhler, A.R.; Hilty, L.M.; Bakker, C. Prospective impacts of electronic textiles on recycling and disposal. *J. Ind. Ecol.* **2011**, *15*, 496–511. [CrossRef]
162. Sun, C.H.; Shang, G.Q.; Tao, Y.Y.; Li, Z.R. A review on application of piezoelectric energy harvesting technology. *Adv. Mater. Res.* **2012**, *516*, 1481–1484. [CrossRef]
163. Can Electricity from the Human Body Replace Batteries? BBC News. Available online: http://www.bbc.co.uk/news/science-environment-19470850 (accessed on 22 August 2017).
164. Burns, M.L. *Medical Trauma Assessment through the Use of Smart Textiles*; Final Technical Report 7/14/94–2/28/95; Science, Math & Engineering, Inc.: Billerica, MA, USA, 1995.

© 2018 by the authors. Licensee MDPI, Basel, Switzerland. This article is an open access article distributed under the terms and conditions of the Creative Commons Attribution (CC BY) license (http://creativecommons.org/licenses/by/4.0/).

Article

Effect of Fabric Integration on the Physical and Optical Performance of Electroluminescent Fibers for Lighted Textile Applications

Alyssa Martin * and Adam Fontecchio

Department of Electrical, Computer Engineering, Drexel University, Philadelphia, PA 19104, USA; fontecchio@coe.drexel.edu or aab65@drexel.edu or af63@drexel.edu
* Correspondence: alyssa.a.martin28@gmail.com; Tel.: +1-(215)-895-3232

Received: 20 April 2018; Accepted: 13 July 2018; Published: 17 July 2018

Abstract: The advent of electroluminescent (EL) fibers, which emit light in response to an applied electric field, has opened the door for fabric-integrated light emission and displays in textiles. However, there have been few technical publications over the past few years about the performance of these light emitting fibers inside functional fabrics. Thus, there is limited information on the effect of integration on the physical and optical performance of such devices. In this work, alternating current powder-based EL (ACPEL) fibers were evaluated under a range of operating conditions both inside and outside of a knit matrix to understand how the EL fiber device performance changed inside a functional fabric. The device efficiency, adjustable brightness, and mechanical properties of these fibers are presented. The effects of fabric integration on the light-emitting fibers as well as the supporting knit fabric are discussed as they relate to the practical applications of this technology.

Keywords: smart fabric; light-emitting textile; electroluminescent fiber; light-emitting fiber; physical characterization; optical characterization

1. Introduction

The merging of electronics and textiles has given rise to garments and upholstery with new abilities such as sensing, biomedical monitoring, power storage, movement, and communication. Within this smart fabric field, light emitting fibers are garnering great attention for applications in fashion, entertainment, optical physiological monitoring, safety lighting in garments, and automotive and aircraft interior lighting [1]. The integration of electronics and textiles has typically been accomplished by mounting prefabricated devices into garments [2,3] and incorporating discrete components (e.g., sensors, batteries, controller chips) with laminated or knit conducting interconnects [4,5] that compromise the most desirable characteristics of the textile including conformability, softness, strength, and washability. Therefore, the smart fabric arena is rapidly pushing towards fabric-integrated electronic systems and fibers and fabric structures exhibiting electro-optic properties [6,7].

A number of light emitting fibers have emerged in the literature including side-emitting optical fibers [8–10], mechanoluminescent fibers [11], electroluminescent (EL) fibers [12–19], and fibers exhibiting photoluminescence [20–23]. Electroluminescent and photoluminescent fibers emit light upon the application of electrical or optical power, respectively. Both optical and photoluminescent (PL) fibers require an external light source for illumination; optical fibers must be illuminated at one end by a luminescent source and PL fibers must be charged by the sun or external luminescent source. Fibers that emit light via electroluminescence are especially desirable for lighted fabrics and fabric-integrated displays due to the fast switching times, inherent luminescence, and adjustable brightness of these electrically controllable devices [24]. EL fibers based on inorganic powder phosphor [17], organic [15], and polymer [12,13,16,18] material systems have been reported in literature. Table 1 compares the optical performance and bending radius of fibers fabricated from different material systems.

Table 1. Comparison of electroluminescent fiber properties reported in literature.

Type of Fiber	Wavelength	Maximum Emission Intensity	Turn On Voltage	Bending Radius	Reference
ACPEL	485 nm	49.39 cd/m^2	50 V	3.1 cm	[17,25]
LEC (iTMC)	855–984 nm	23 cd/m^2	4.2 V	6 mm	[13]
OLED	N/A	104 mW/cm^2	~0.5 V	N/A	[15]
pLEC	N/A	125 cd/m^2	4.2 V	N/A	[12]
PLED (MEH-PPV)	~575 nm	N/A	15–16 V	N/A	[18]
PLED (SY solution)	~550 nm	1458.8 cd/m^2	5 V	2.5 mm	[16]

Despite the higher efficiency and flexibility of some of these other fibers, complicated fiber fabrication processes and the sensitivity of the resulting fibers to handling and environmental factors like heat, moisture, and humidity have prevented many of the reported EL fibers from achieving full fabric-integration [12,13,16,24,26]. The only fibers to be integrated into a knit fabric are those fabricated by Dias et al. [17], which were inlaid into the knit fabric, and Coyle et al. [18], which required the fibers to be placed under tension while testing to maintain contact between the two fibers acting as the top and bottom electrodes in the electroluminescent structure. The fibers produced by Zhang et al. [12] and Kwon et al. [27] produced light while bent, but were not placed inside a fabric structure for testing. A more detailed comparison of fibers produced by these material systems exists in previous literature [25]. The ACPEL material set has demonstrated reliable light emission over large surface areas in planar films and coated fibers due to the robustness of the thick layers [28]. This materials system was chosen for testing due to its simplicity of fabrication, robustness of the thick layers, long lifetimes, and brightness visible to the naked eye.

However, a lot is still unknown about the effect fabric integration has on the physical and optical performance characteristics of these fibers. In this work, the optical, electrical, and mechanical performance of ACPEL fibers inside and outside of a knit fabric were compared to understand the effect of integration on both the light-emitting fibers and knit fabric. The flexibility, robustness, brightness, and power requirements of the light-emitting fabric resulting from the integration of ACPEL fibers into a knit fabric are important in identifying the applications and limitations of the system. These properties and the methods of obtaining them for EL fibers inside and outside of a fabric are reported in this work. These methods provide a foundation for experimentally quantifying and comparing the performance of fabric-integrated EL fibers.

2. Materials and Methods

2.1. Fiber Fabrication

Electroluminescent (EL) fibers were fabricated by coating an alternating current phosphor-based EL (ACPEL) structure onto a supporting conductive fiber. The cross-sectional ACPEL device structure, depicted in Figure 1, consisted of a conductive bottom electrode, an isolation layer with a high dielectric constant to focus the electric field on the emitting layer and protect it from heating at this electrode, an emitting layer, and a top electrode layer that was translucent to allow light through. The Dupont Luxprint® material system was used due to its robustness, simplicity of deposition, and ease of handling after curing. The solution-processed layers were deposited in the following order onto an Aracon® XS0400E-018 silver-coated Kevlar® yarn (manufactured by Micro-coax, Inc., Pottstown, PA, USA): Dupont Luxprint® 8153 dielectric paste, Dupont Luxprint® 8154L phosphor paste, and Dupont Luxprint® 7164 transparent conductive paste. Upon application of the first isolation layer, the dielectric paste filled the gaps between the individual silver coated fibers within the underlying conductive yarn, creating a monofilament onto which subsequent layers were deposited. The individual strands of the supporting conductive yarn were randomly dispersed in the dielectric coating layer.

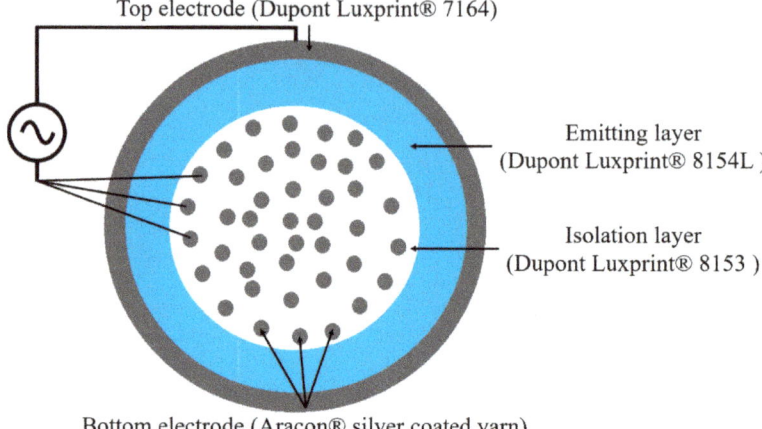

Figure 1. Schematic diagrams of the ACPEL fiber structure.

Slot die devices, like that shown in Figure 2, with varying cylindrical die openings were used to control the deposition of each layer onto the previous layer of the device structure. These devices were described in detail in previous work [29]. The fibers were cured inside an oven to evaporate the solvent from the fluid coating material after deposition. The dielectric and phosphor pastes were cured at 130 °C for 15 min, while the translucent conductive paste took only 5 min to cure at the same temperature [30]. Unlike many other organic and polymer emissive systems [24], this material set is not highly sensitive to surface roughness, oxygen, and particulates from the air due to its thick layers. Therefore, the entire fabrication process can take place in air and outside a clean room, which is a great advantage for scaling production of the fibers and using them in commercial applications.

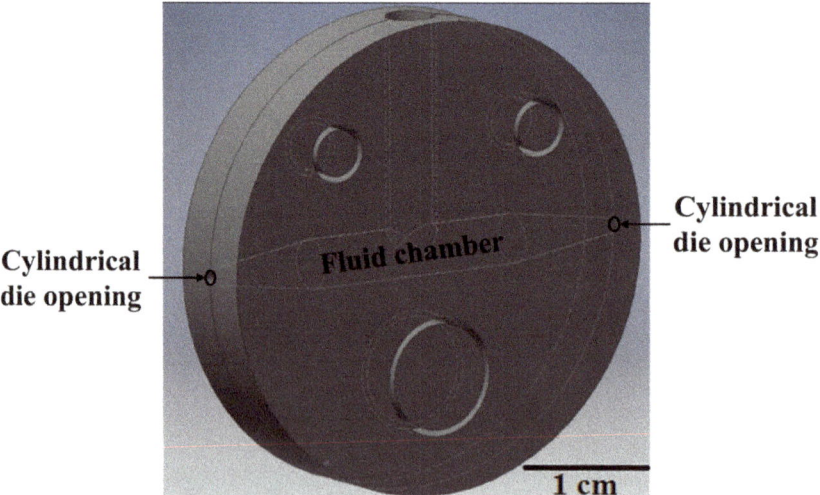

Figure 2. Model of slot die devices used to deposit ACPEL layers onto supporting yarn substrate [29].

2.2. Power Application and Fabric Integration

An Agilent 33220A arbitrary waveform generator (AWG) amplified by a TREK model PZD700 high voltage amplifier supplied power to the fibers. Power was applied between the top electrode and across all individual fibers of the supporting conductive yarn electrode simultaneously to illuminate the fiber. This was accomplished by twisting the individual fibers of the supporting conductive yarn together and contacting them using flat alligator clips as shown in Figure 3a. The more tightly packed the individual fibers of the supporting yarn electrode, the more efficient the devices because there is less power lost to the dielectric isolation layer. However, as long as the average isolation layer thickness between the emissive layer and conductive fibers of the supporting yarn closest to that emissive layer has a 35 ± 10 µm tolerance, light emission across the yarn will appear uniform to an observer. This thickness is controlled by the diameter of the cylindrical opening of the slot die coating device.

The illuminated fiber in Figure 3b had non-uniform emission due to an uneven dispersion of supporting conductive fibers. The thicknesses of the layers has been reported in previous work [29]. Although the total thickness of the isolation layer was consistent across, the parts of the fiber that appear dark had an isolation layer thickness between the emissive layer and conductive fibers of the supporting yarn greater than 50 µm. These dark areas were eliminated by keeping the fibers of the supporting yarn tightly twisted using clamps during deposition and curing to control the dispersion of the fibers in the isolation layer.

A 3 cm × 3 cm weft knit fabric composed of a 3-ply cotton yarn (Size 10 crochet cotton thread, Coats & Clark, Inc., Charlotte, NC, USA) with a gauge of 4 rows and 3 stitches = 1 cm was hand knit. A single fiber was inlaid into the fabric as shown in Figure 3c. The conductive ends of the fibers were accessible on either end of the fabric sample, which allowed power to be applied using flat alligator clips.

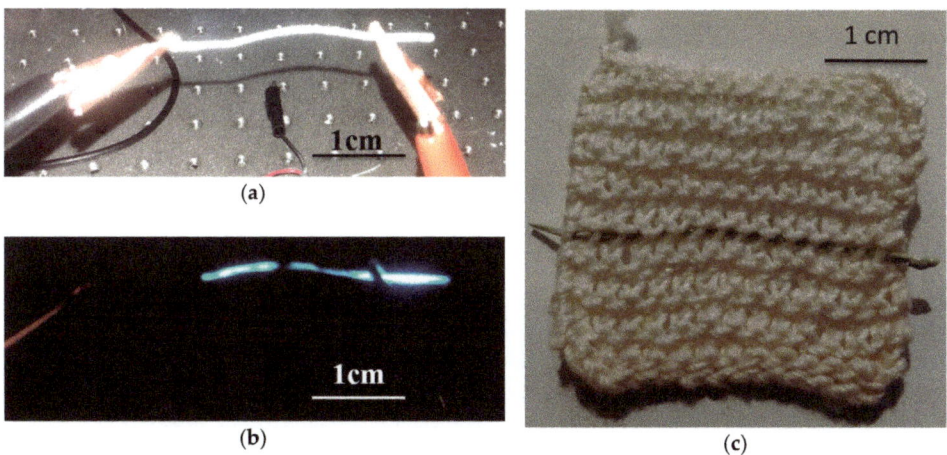

Figure 3. Macro images of the ACPEL fibers (**a**) outside of knit indicating where electrodes are connected to fiber during testing, (**b**) illuminated at 100 V$_{rms}$, 400 Hz, and (**c**) inlaid into 3 cm 2 knit fabric sample.

2.3. Optical Characterization Methods

Electrical, optical, mechanical tests were performed on EL fibers inside and outside of the cotton knit matrix. For each test, a 3-cm segment of fiber was tested due to space constraints in the fabrication and testing set ups. However, using a longer continuous segment of fiber or multiple fibers would increase the light output and effect the mechanical strength and flexibility of the light-emitting fabric.

Optical testing was performed inside a 12-inch diameter Gamma Scientific integrating sphere to eliminate any geometric and ambient light effects on the intensity of light. The intensity of the emitted light was measured by an Ocean Optics USB-4000 spectrometer attached via fiber optic cable to the integrating sphere and Spectrasuite software was used to process the data. Prior to beginning each optical measurement, a voltage waveform with amplitude below the threshold voltage of the device was applied to the fiber for half an hour to allow the device to reach the steady-state.

The voltage, frequency, and shape of the applied waveform can affect the optical output of the device. ACPEL devices are most efficient during the rising edge portion of the applied waveform, and least efficient when the voltage is held constant above the threshold voltage. A sine wave was used to illuminate the ACPEL fiber devices except for the device efficiency measurements. The threshold voltage of the EL fiber device was 49.833 V, which is independent of frequency. This voltage is the magnitude of power necessary to accelerate electrons inside the EL layer to speeds high enough to produce light via impact ionization with phosphor particles in the emitting layer.

The frequency of the applied waveform plays a large role in the brightness of light emission in inorganic phosphor devices. The field must switch fast enough to have continuous excitation and decay of electrons, but cannot exceed the lifetime of electrons. The brightness-frequency (B-F) curve was determined by applying waveforms with increasing frequency to the device, while keeping the amplitude constant, and measuring the luminance at each increased frequency. The frequency was increased in increments of 100 Hz from 0 Hz to 50 kHz, while the voltage remained constant at 100 V. A constant voltage of 100 V was selected for the frequency sweep as it is in the middle of the fiber operating range. The brightness-voltage (B-V) curve was determined by applying waveforms with increasing amplitude (higher voltage) at a constant frequency of 400 Hz to the device and measuring the luminance at each successive amplitude. The voltage sweep was performed on fibers at a constant frequency of 400 Hz, while the voltage was increased in increments of 10 V from 0 V to 230 V. To determine if the knitted matrix absorbed or blocked any light from the fibers, voltage and frequency sweeps were performed on fibers inlaid into and outside a knitted fabric matrix.

2.4. Device Efficiency Calculation

A bipolar trapezoidal waveform, shown in Figure 4, was used to drive the fiber in the circuit as there are distinct points on the waveform that can be identified and plotted to obtain a Q-V curve. These points have been explained in great detail in prior work [25,31]. Most of the points designated where the applied voltage began increasing and decreasing, except for B and G, which represent the points in the waveform where the device began conducting charge, commonly called the turn-on voltage.

The device efficiency can be derived from the charge-voltage (Q-V) curve of the ACPEL fibers. The density of electrical power delivered across the ACPEL device per pulse is equal to the area inside the Q-V curve. This measurement divided by the input power equals the device efficiency. The Sawyer-Tower method is commonly used to determine this curve in EL devices [32,33]. In this method, the voltage drop across a sense element, typically a capacitor or resistor, is monitored to gain information about device operation. The Sawyer-Tower circuit is depicted in Figure 5. The signal applied to the circuit consisted of a 1 kHz sequence of bipolar pulses with 100 V amplitude, rise, and fall times of 5 µs and a pulse width of 30 µs.

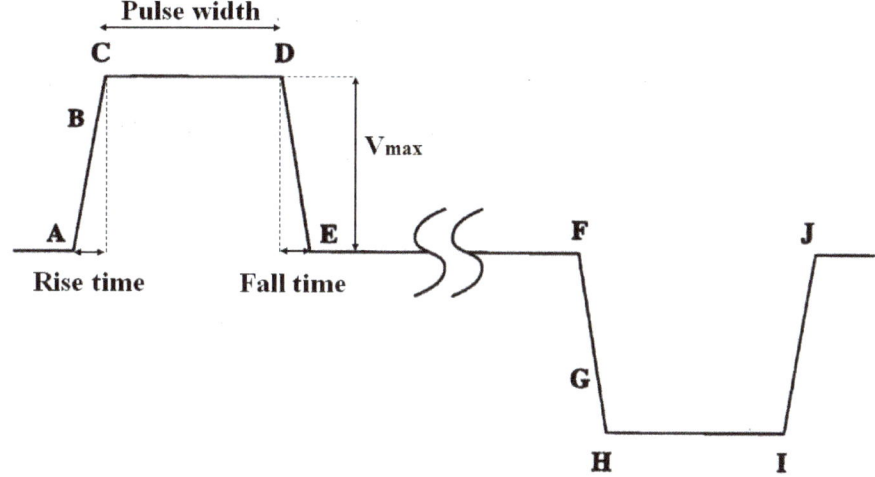

Figure 4. Bipolar trapezoid waveform indicating important measurement points.

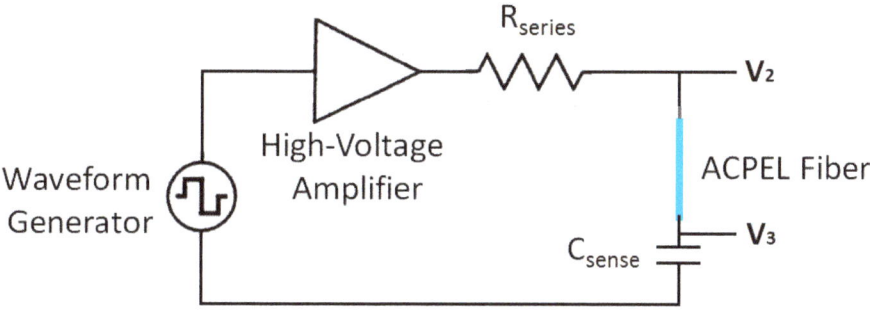

Figure 5. Schematic diagram of Sawyer-Tower measurement circuit set up.

The output of the high-voltage amplifier drives the characterization circuit, which is a series combination of a resistor, the ACPEL fiber device, and a sense element. The sense element in this case was a capacitor, which was chosen to be much larger than the capacitance of the ACTFEL device to minimize its effect on the measurement. The resistor acts as a current-limiter to protect the ACPEL fiber device from catastrophic failure. The capacitance of a pixel in the 3 cm segment of ACPEL fiber was measured with a GW Instek LCR-819 LCR meter as 0.27 nF, so an 82 nF capacitor was used as the sense element (C_{sense}) in the characterization circuit. The resistor acts as a current-limiter to protect the ACPEL fiber device from catastrophic failure, so a 1.5 kΩ resistor was used. The voltages V_2 and V_3 in Figures 6 and 7 were monitored by an Agilent MSO-X 2014A oscilloscope during testing.

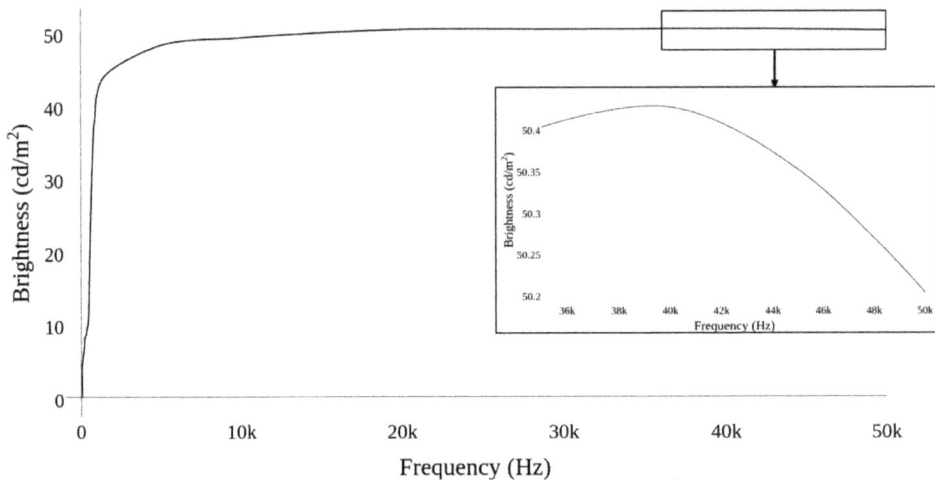

Figure 6. Effect of frequency on the luminance of the inlaid fiber (Voltage = 100 VAC).

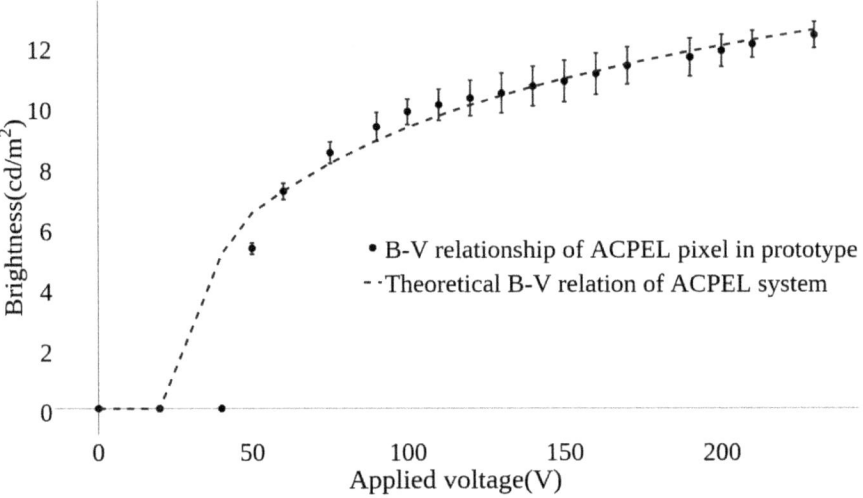

Figure 7. Comparison of the theoretically and experimentally derived relationship between brightness and voltage of an ACPEL fiber inside a knit fabric.

The voltage drop across the sense capacitor, $V_3(t)$, is related to the amount of charge stored on the external terminals of the ACPEL fiber device, $q_{ext}(t)$, and described by

$$q_{ext}(t) = C_{sense} V_3(t) \tag{1}$$

The value of $q_{ext}(t)$ is also called the instantaneous external charge because it is a transient charge measured externally with respect to the phosphor of the ACPEL device. The Q-V characteristics are obtained by plotting this instantaneous electric charge against the instantaneous voltage drop across the ACPEL fiber device, $V_i(t)$, which is determined by

$$V_i(t) = V_2(t) - V_3(t) \tag{2}$$

2.5. Mechanical Characterization Methods

The force vs. displacement curves of the EL fibers inside and outside of the knitted matrix and the bending radius of the fibers were previously reported [34] and demonstrated that the fibers were not flexible enough to be used in garments. To further explain the effect in both the knitted fabric and EL fiber after integration and determine what applications the fibers can be used for, flexural stress-strain curves were derived from the three-point flexural test detailed by ASTM D6856/D6856M [35,36]. Experimental tests were performed with a Mark-10 Force Gauge model MS-20 on fibers inside and outside of a fabric matrix. Details of this testing set up are given in a previous work [34].

Equations (3) and (4) are based on homogeneous beam theory and define the flexural stress for a circular cross section (σ_f) and flexural strain (ε_f) on the outer surface of the fiber, respectively [36]. In these equations, F is applied load, L is the support span, R is the radius of the fiber, D is the deflection of the fiber, and d is the diameter of the beam [37].

$$\sigma_f = \frac{FL}{\pi R^3} \quad (3)$$

$$\varepsilon_f = \frac{6Dd}{L^2} \quad (4)$$

These equations were used to calculate the flexural stress-strain response of fibers inside and outside of the knit fabric as an increasing load was applied. The three-point bend test method is sensitive to the geometry of the fiber, and the accuracy of the calculated values depends upon the ratio of the beam length (L) to the cross-sectional height of the fiber (h), called the span to thickness ratio (L/h). A L/h ratio of at least 20:1 is recommended for determining the flexural modulus of fiber-reinforced composites [36]. During experimentation, the L/h ratio was 32:1, which is in the recommended range of values for determining the flexural modulus according to ASTM D6856/D6856M.28 [35].

3. Results

3.1. Optical Characterization

Figure 6 shows the results of the frequency sweep. The intensity of light increased with increasing frequency until approximately 45 kHz, when darkening at the center began to appear and the overall intensity decreased. This darkening was a result of the increase in degradation of the electroluminescent devices under very high frequencies. There is minimal gain in luminance with a frequency above 5 kHz and continuous operation of the fiber devices at higher frequencies will decrease the lifetime of the fiber, so it is better to run the device at frequencies below this point [30].

The relationship between brightness and voltage is well defined in the literature for ACPEL material systems and is described by Equation (5), where U is the applied voltage and A and c are empirical constants [38,39]. These empirical constants are dependent on the material properties of the ACPEL system, and are unique to the material under investigation. Once the empirical constant has been derived, the brightness of the fiber at different voltages can be theoretically predicted for this fiber. Alternatively, one can predict what voltage would be needed to achieve a desired brightness from the material and determine if these fibers will fit into an application based on the limitations of the device.

$$B = A exp\left(-\frac{c}{U^{\frac{1}{2}}}\right) \quad (5)$$

The empirical constants of the ACPEL material system used in this work were derived using a nonlinear regression analysis to fit the equation to the average of five experimental voltage sweeps performed on a single pixel in the display. Figure 7 demonstrates the accuracy of the brightness-voltage (B-V) curve predicted by Equation (5), where the empirical constants A and c for this ACPEL system were calculated as 22.2678 and 8.7396, respectively. The error bars at each voltage increment of the experimentally derived curve show the standard deviation to the average of the five voltage sweeps performed on the pixel. The difference in luminance of a fiber inside and outside of a knitted fabric matrix was measured as 0.00015 ± 0.00005 cd/m^2 regardless of the voltage or frequency applied.

3.2. Device Efficiency

Figure 8 depicts the Q-V curve of an EL fiber at 100 V. The density of power delivered to the device per pulse was 323.64 W/cm^3. The power efficiency of the device could then be derived by dividing this delivered power by the input power, which gave an efficiency of 0.016% for this device at 40 V above threshold. This efficiency depends upon a number of factors including the concentration of phosphor particles in the emitting layer, applied waveform, length of the fiber, environmental factors, and the dielectric constant of both the isolation layer and suspension medium of the emitting layer [31]. Thus, the efficiency of the device can vary over time and under different operating conditions.

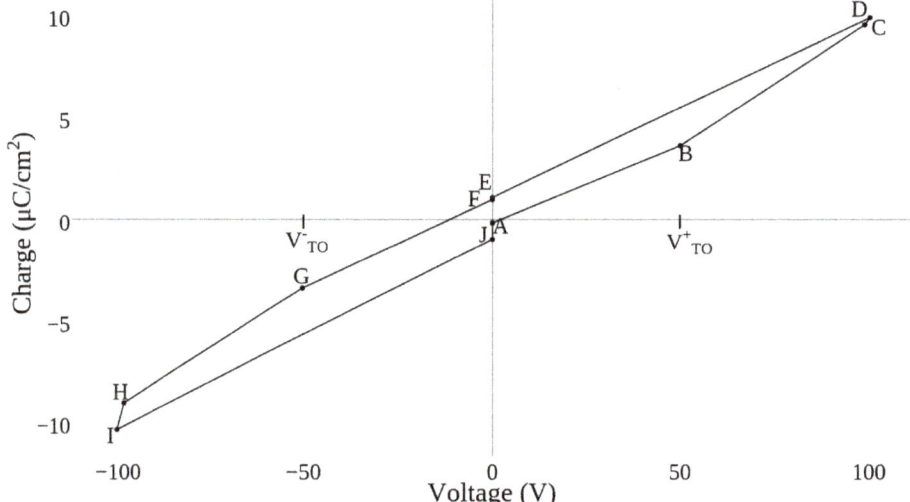

Figure 8. Charge-Voltage (Q-V) curve of the 3 cm ACPEL fiber at 100 V.

3.3. Mechanical Characterization

The curves resulting from the mechanical tests, shown in Figure 9, indicated that the knitted matrix absorbed some of the applied stress and allowed the inlaid fiber to withstand a much higher load than the fiber on its own. The flexural strength is a measure of how much force or pressure is needed to break the fibers, which is quantified by the maximum stress before fiber failure. Assuming that the maximum stress occurs at the outermost layer in the EL fiber, the flexural strength can be calculated by Equation (3), where F is the maximum force applied before failure [40]. The flexural strength of the fibers inside and outside the knitted matrix was calculated as 5.23 MPa and 2.02 MPa, respectively. Based on these values, the knitted matrix allows the fiber to withstand more than twice the load that it would be able to on its own.

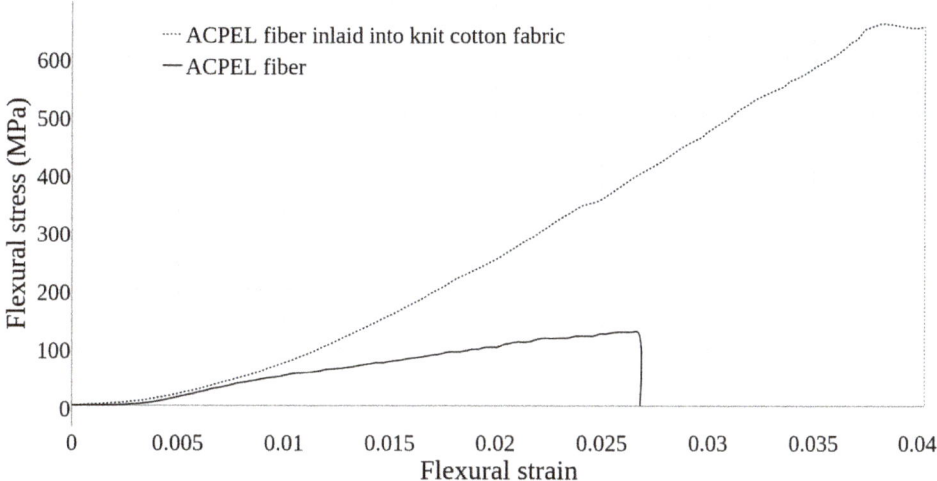

Figure 9. Bipolar trapezoid waveform indicating important measurement points.

The flexural modulus of the fibers is the ratio of stress to strain in flexural deformation, which describes the material's tendency to bend. The higher the flexural modulus, the more resistant it is to bending. Ideally, this flexural modulus is equivalent to the elastic modulus of the material, but in reality, these values may differ due to the presence of normal and shear stresses throughout the beam span. The contribution of the shear stress can be reduced by using a high L/h ratio. During experimentation, a sufficiently high L/h ratio was maintained to minimize this effect. The flexural modulus of an EL fiber outside the knit textile was determined using Equation (6), where m is the initial slope of the load-deflection curve given in Figure 9 [41].

$$E_f = \frac{FL^3}{48\pi R^4 D} \quad (6)$$

The flexural modulus of a single fiber was determined to be 4.9 GPa, which means that these fibers are stiff and have a high resistance to bending.

4. Discussion

The frequency has a more pronounced effect on the brightness of the ACPEL system than the voltage because of the time dependence of light emission in ACPEL devices. Upon application of an electric field with a high enough magnitude, electrons are excited by impact ionization and emit light as they relax back to ground state. This relaxation process occurs on the order of nanoseconds, and the emission produced has an equivalent decay time. To have a continuous emission, the applied field must be constantly switched as the emission and subsequent decay of the devices occurs twice per AC cycle. Generally, a higher frequency will result in a higher brightness due to the time dependence of the emission [28]. However, once the AC cycle becomes comparable to or surpasses the lifetime of excited electrons, the brightness will decrease.

A number of textile properties such as gauge, yarn thickness, and yarn material could affect the visible light output of the fibers. Specifically, a tighter gauge, thicker yarn, or denser yarn material could decrease the amount of visible light. The apparent brightness of the fibers will also change based on the amount of ambient light. In applications where the surroundings are dark, like a wall mounted screen in a movie theater or night time concert, the fibers could be run at lower voltages and frequencies, thus increasing the lifetime of the device.

The efficiency of the fiber is an internal property of the fiber that will not change with fabric integration. However, the wear and tear that the fiber experiences during normal use in a functional fabric will decrease this efficiency over time. The efficiency is an important property of the fiber as it determines the power requirements of the fiber. The low device efficiency exhibited by the ACPEL fibers is a consequence of the indirect method of light emission in these devices, which gives rise to high power requirements. However, this AC power can be supplied directly in airplanes and through wall electrical outlets, enabling automotive interior lighting and wall-mounted lighting applications.

Moisture, and therefore humidity, can accelerate the natural decrease in luminous efficiency that occurs over time in inorganic phosphor devices [42]. The transparent conductive top layer reduces the exposure of the phosphor layer to such environmental effects. However, fabrics incorporating these fibers should not be subjected to excessive moisture like soaking or washing without additional precautions to protect the fibers. Additionally, the high power requirements significantly reduce the portability of the fibers and fabrics supporting these fibers and require that the fibers be isolated from human skin. In the future, an encapsulation layer could be added to protect the fibers and electrically isolate them to prevent human contact. However, the addition of more layers could further reduce the flexibility of the fibers, so a highly flexible polymer encapsulation layer is recommended.

The ACPEL fibers exhibit a high resistance to bending, which has been discussed in previous work [29,34]. Thus, when introduced into the knitted matrix, they reduced the overall flexibility, but significantly increased the strength of the fabric. This increased strength results from the distribution of force across the fiber and matrix. The fiber absorbs some of the force and as it has a high flexural strength, it reinforces the fabric strength. This increased strength is useful in applications where the fibers is subjected to a great deal of perpendicular stresses such as automotive carpet lighting or a display screen directly integrated into the fabric of a car seat.

5. Conclusions

In this paper, ACPEL fiber devices were fabricated and experimentally analyzed. To demonstrate the performance and potential of these fibers in functional fabrics, electrical, optical, and mechanical tests were performed on the fibers inside and outside of a knit fabric. The simple fabrication procedures and ease of handling streamlines the integration of ACPEL fibers into knitted fabrics. The integration of ACPEL fibers into a knit structure improves the strength and robustness of the resulting light emitting fabric, making it useful for applications in automotive/airplane interior lighting and emissive wall displays.

Author Contributions: A.M. fabricated the EL devices, performed the characterization experiments, analyzed the data and wrote the paper. A.F. contributed all other materials/analysis tools and revised the manuscript.

Funding: This material is based upon work supported by both the National Science Foundation Graduate Research Fellowship under Grant No. DGE-1104459, and the National Science Foundation STEM K-12 Fellowship under grant No. DGE-0947936.

Conflicts of Interest: The authors declare no conflict of interest.

References

1. Dalsgaard, C.; Sterrett, R. White paper on smart textile garments and devices: A market overview of smart textile wearable technologies. In *Market Opportunities for Smart Textiles*; Ohmatex: Viby, Denmark, 2014.
2. Janietz, S.; Gruber, B.; Schattauer, S.; Schulze, K. Integration of OLEDs in textiles. *Adv. Sci. Technol.* **2013**, *80*, 14–21. [CrossRef]
3. Wang, G.F.; Tao, X.M.; Huang, H.M. Light-emitting devices for wearable flexible displays. *Color. Technol.* **2005**, *121*, 132–138. [CrossRef]
4. Orth, M.; Post, R.; Cooper, E. Fabric computing interfaces. In *CHI 98 Conference Summary on Human Factors in Computing Systems—CHI '98*; ACM Press: New York, NY, USA, 1998; pp. 331–332.
5. CuteCircuit. Available online: https://cutecircuit.com/ (accessed on 8 July 2016).

6. Mauriello, M.; Gubbels, M.; Froehlich, J. Social fabric fitness: The design and evaluation of wearable E-textile displays to support group running. In Proceedings of the SIGCHI Conference on Human Factors in Computing Systems, Toronto, ON, Canada, 26 April–1 May 2014.
7. Jost, K.; Stenger, D.; Perez, C.R.; McDonough, J.K.; Lian, K.; Gogotsi, Y.; Dion, G. Knitted and screen printed carbon-fiber supercapacitors for applications in wearable electronics. *Energy Environ. Sci.* **2013**, *6*, 2698–2705. [CrossRef]
8. Koncar, V. Optical fiber fabric displays. *Opt. Photonics News* **2005**, *16*, 40–44. [CrossRef]
9. Spigulis, J. Side-emitting fibers brighten our world. *Opt. Photonics News* **2005**, *16*, 34–39. [CrossRef]
10. Deflin, E.; Koncar, V.; Weill, A.; Vinchon, H. Bright optical fiber fabric: A new flexible display. *Text. Asia* **2001**, *33*, 25–28.
11. Jeong, S.M.; Song, S.; Seo, H.-J.; Choi, W.M.; Hwang, S.-H.; Lee, S.G.; Lim, S.K. Battery-Free, Human-Motion-Powered Light-Emitting Fabric: Mechanoluminescent Textile. *Adv. Sustain. Syst.* **2017**, *1*, 1700126. [CrossRef]
12. Zhang, Z.; Guo, K.; Li, Y.; Li, X.; Guan, G.; Li, H.; Luo, Y.; Zhao, F.; Zhang, Q.; Wei, B.; et al. A colour-tunable, weavable fibre-shaped polymer light-emitting electrochemical cell. *Nature* **2015**, *9*, 233–238. [CrossRef]
13. Yang, H.; Lightner, C.R.; Dong, L. Light-emitting coaxial nanofibers. *ACS Nano* **2012**, *6*, 622–628. [CrossRef] [PubMed]
14. Samei, E.; Badano, A.; Chakraborty, D.; Compton, K.; Cornelius, C.; Corrigan, K.; Flynn, M.J.; Hemminger, B.; Hangiandreou, N.; Johnson, J.; et al. Assessment of display performance for medical imaging systems. *Med. Phys.* **2005**, *32*, 1205–1225. [CrossRef] [PubMed]
15. O'Connor, B.; An, K.K.H.H.; Zhao, Y.; Pipe, K.P.P.K.; Shtein, M. Fiber Shaped Light Emitting Device. *Adv. Mater.* **2007**, *19*, 3897–3900. [CrossRef]
16. Kwon, S.; Kim, W.; Kim, H.C.; Choi, S.; Park, B.-C.; Kang, S.-H.; Choi, K.C. P-148: Polymer Light-Emitting Diodes Using the Dip Coating Method on Flexible Fiber Substrates for Wearable Displays. In Proceedings of the 53rd Annual SID Symposium, San Jose, CA, USA, 21 May–5 June 2015; Volume 46, pp. 1753–1755.
17. Dias, T.; Monaragala, R. Development and analysis of novel electroluminescent yarns and fabrics for localized automotive interior illumination. *Text. Res. J.* **2012**, *82*, 1164–1176. [CrossRef]
18. Coyle, J.P.; Li, B.; Dion, G.; Fontecchio, A.K. Direct integration of a 4-pixel emissive display into a knit fabric matrix. *Adv. Disp. Technol. III* **2013**, *8643*, 864308. [CrossRef]
19. Zhang, Z.; Cui, L.; Shi, X.; Tian, X.; Wang, D.; Gu, C.; Chen, E.; Cheng, X.; Xu, Y.; Hu, Y.; et al. Textile Display for Electronic and Brain-Interfaced Communications. *Adv. Mater.* **2018**, *30*, 1800323. [CrossRef] [PubMed]
20. Services, S.A.T. Glow Yarns for Special Effects in Textiles from Swicofil. Available online: http://www.swicofil.com/glow_yarn.html (accessed on 23 June 2017).
21. Bortz, T.; Agrawal, S.; Shelnut, J. Photoluminescent Fibers, Compositions and Fabrics Made Therefrom. U.S. Patent 8,207,511, 26 June 2012.
22. Gauvreau, B.; Guo, N.; Schicker, K.; Stoeffler, K.; Boismenu, F.; Ajji, A.; Wingfield, R.; Dubois, C.; Skorobogatiy, M. Color-changing and color-tunable photonic bandgap fiber textiles. *Opt. Express* **2008**, *16*, 15677–15693. [CrossRef] [PubMed]
23. Sayed, I.; Berzowska, J.; Skorobogatiy, M. Jacquard-woven photonic bandgap fiber displays. *Res. J. Text. Appar.* **2010**, *14*, 97–105. [CrossRef]
24. O'Connor, B.; Pipe, K.; Shtein, M. Fiber based organic photovoltaic devices. *Appl. Phys. Lett.* **2008**, *92*, 172. [CrossRef]
25. Bellingham, A. *Direct Integration of Dynamic Emissive Displays into Knitted Fabric Structures*; Drexel University: Philadelphia, PA, USA, 2017.
26. Kim, W.; Kwon, S.; Lee, S.-M.; Kim, J.Y.; Han, Y.; Kim, E.; Choi, K.C.; Park, S.; Park, B.-C. Soft fabric-based flexible organic light-emitting diodes. *Org. Electron.* **2013**, *14*, 3007–3013. [CrossRef]
27. Kwon, S.; Kim, W.; Kim, H.; Choi, S.; Park, B.-C.; Kang, S.-H.; Choi, K.C. High Luminance Fiber-Based Polymer Light-Emitting Devices by a Dip-Coating Method. *Adv. Electron. Mater.* **2015**, *1*, 1500103. [CrossRef]
28. Bredol, M.; Dieckhoff, H.S. Materials for Powder-Based AC-Electroluminescence. *Materials* **2010**, *3*, 1353–1374. [CrossRef]
29. Bellingham, A.; Bromhead, N.; Fontecchio, A. Rapid Prototyping of Slot Die Devices for Roll to Roll Production of EL Fibers. *Materials* **2017**, *10*, 594. [CrossRef] [PubMed]

30. Dupont. Processing Guide for DuPont Luxprint Electroluminescent Inks. 2012. Available online: http://www.dupont.com/content/dam/dupont/products-and-services/electronic-and-electricalmaterials/documents/prodlib/MCM-EL-Processing-Guide.pdf (accessed on 16 July 2018).
31. Keir, P. Fabrication and Characterization of ACTFEL Devices, Oregon State. 1999. Available online: https://ir.library.oregonstate.edu/concern/graduate_thesis_or_dissertations/n009w490c (accessed on 16 July 2018).
32. Aleksandrova, M.P.; Dobrikov, G.H.; Andreev, S.K.; Dobrikov, G.M.; Rassovska, M.M. Electrical Properties Characterization of Thick Film Organic Electroluminescent Structures. *Annu. J. Electron.* **2011**, *5*, 183–186.
33. Wager, J.F.; Keir, P.D. Electrical Characterization of thin-film electroluminescent devices. *Annu. Rev. Mater. Sci.* **1997**, *27*, 223–248. [CrossRef]
34. Bellingham, A.; Fontecchio, A. Direct integration of individually controlled emissive pixels into knit fabric for fabric-based dynamic display. *IEEE Photonics J.* **2017**, *9*, 1–10. [CrossRef]
35. Standard Guide for Testing Fabric-Reinforced Textile Composite Materials. ASTM International: West Conshohocken, PA, USA. Available online: http://www.astm.org/cgi-bin/resolver.cgi?D6856D6856M (accessed on 16 July 2018).
36. Zweben, C.; Smith, W.; Wardle, M. Test methods for fiber tensile strength, composite flexural modulus, and properties of fabric-reinforced laminates. In *Composite Materials: Testing and Design (Fifth Conference)*; ASTM International: West Conshohocken, PA, USA, 1979.
37. Oberg, E.; Jones, F.; Horton, H.; Ryffel, H. *Machinery's Handbook*; Industrial Press: New York, NY, USA, 2004.
38. Zalm, P.; Diemer, G.; Klasens, H. Some aspects of the voltage and frequency dependence of electroluminescent zinc sulphide. *Philips Res. Rep.* **1955**, *10*, 205–215.
39. Alfrey, G.F.; Taylor, J.B. Electroluminescence in Single Crystals of Zinc Sulphide. *Proc. Phys. Soc. Sect. B* **1955**, *68*, 775–784. [CrossRef]
40. Hodgkinson, J.M. *Mechanical Testing of Advanced Fibre Composites*; Elsevier: New York, NY, USA, 2000; ISBN 9781855738911.
41. Mallick, P.K. *Fiber-Reinforced Composites: Materials, Manufacturing, and Design*; CRC/Taylor & Francis: Boca Raton, FL, USA, 2007; ISBN 1420005987.
42. Smet, P.F.; Moreels, I.; Hens, Z.; Poelman, D. Luminescence in Sulfides: A Rich History and a Bright Future. *Materials* **2010**, *3*, 2834–2883. [CrossRef]

© 2018 by the authors. Licensee MDPI, Basel, Switzerland. This article is an open access article distributed under the terms and conditions of the Creative Commons Attribution (CC BY) license (http://creativecommons.org/licenses/by/4.0/).

Article

A Novel Method for Embedding Semiconductor Dies within Textile Yarn to Create Electronic Textiles

Mohamad-Nour Nashed, Dorothy Anne Hardy *, Theodore Hughes-Riley * and Tilak Dias

Advanced Textiles Research Group, Nottingham Trent University, Nottingham NG1 4FQ, UK; m-nour.nashed2014@my.ntu.ac.uk (M.-N.N.); tilak.dias@ntu.ac.uk (T.D.)
* Correspondence: dorothy.hardy@ntu.ac.uk (D.A.H.); theodore.hughesriley@ntu.ac.uk (T.H.-R.); Tel.: +44-115-84-82709 (D.A.H.); +44-0-115-848-8178 (T.H.-R.)

Received: 21 December 2018; Accepted: 21 January 2019; Published: 26 January 2019

Abstract: Electronic yarns (E-yarns) contain electronics fully incorporated into the yarn's structure prior to textile or garment production. They consist of a conductive core made from a flexible, multi-strand copper wire onto which semiconductor dies or MEMS (microelectromechanical systems) are soldered. The device and solder joints are then encapsulated within a resin micro-pod, which is subsequently surrounded by a textile sheath, which also covers the copper wires. The encapsulation of semiconductor dies or MEMS devices within the resin polymer micro-pod is a critical component of the fabrication process, as the micro-pod protects the dies from mechanical and chemical stresses, and hermetically seals the device, which makes the E-yarn washable. The process of manufacturing E-yarns requires automation to increase production speeds and to ensure consistency of the micro-pod structure. The design and development of a semi-automated encapsulation unit used to fabricate the micro-pods is presented here. The micro-pods were made from a ultra-violet (UV) curable polymer resin. This work details the choice of machinery and methods to create a semi-automated encapsulation system in which incoming dies were detected then covered in resin micro-pods. The system detected incoming 0402 metric package dies with an accuracy of 87 to 98%.

Keywords: electronics packaging; encapsulation; electronic yarns (E-yarn); textiles; electronic textiles (E-textiles); smart textiles; intelligent textiles; UV curing; polymer resin

---

1. Introduction

Clothing consists of textile fibers, which are either woven or knitted to produce fabric for both protective and aesthetic purposes. Interactive textiles add a third dimension to traditional textiles [1], with the inclusion of interactivity often achieved using electronics [2]. Interest in the integration of electronics into textiles has increased significantly in recent years, leading to a number of innovative textiles with integrated lighting [3], computing [4], and sensing capabilities [5] for a variety of applications. Electronic textiles can also be used to generate and store electricity [6]. A recent review of the history of electronic textiles is available elsewhere [7], which gives details of a wide variety of electronic textile products that have been designed and developed.

Electronics can be integrated into textiles in one of three ways that can be described as generations of electronic textiles. In the first generation of electronic textiles, electronic components were mounted directly onto the surface of garments. The electronics did not form a part of the structure of the textiles. Examples include sports bras with heart-rate monitors that are attached to the garment with snap fasteners [8]. Thin layers with electronic functionality can be printed directly onto the surface of flexible substrates that include textiles [9]. Deposition processes can also be used to add electronic functionality to the surface of textiles, for example, in the production of flexible photovoltaics [10]. The second generation involved integrating electronics into the structure of garments through knitting, weaving, and embroidery to add electronic functionality. Most focused on the use of conductive fibers to create electrodes or conductive pathways, such as pressure sensor fabrics [11], fabric transducers [12], and a Wearable Motherboard™ [13].

The third generation of electronic textiles could be described as yarns into which sensors and other electronic components were incorporated. E-textile development projects in which electronic functionality has been incorporated at a yarn level include:

- A recent innovation by MIT in which LEDs (light-emitting diodes) and photodiodes were incorporated into a fiber as part of an extrusion process [14]. The polymer extrusion process is likely to result in a yarn with poor tensile strength [15] due to the inability to draw the filaments soon after extrusion without damaging the copper wire interconnects. This will influence the processability of the yarns using standard textile fabrication processes, such as knitting. The MIT concept is also limited to two-terminal devices, which will restrict the range of devices and the scope of functionality that can be incorporated into these yarns;
- The European Union funded PASTA project in which E-Thread® was developed. This saw a die connected to two conducting interconnects, and the die and interconnects covered in a fibrous cover [16]. These dies were not protected by encapsulation, and the E-Thread® could not be subjected to washing processes. The E-Thread® could only be constructed using two-terminal devices, limiting the range of devices, and therefore functions, that can be incorporated. E-thread containing RFID (radio frequency identification) devices are produced by Primo1D [17];
- the Wearable Computing Lab at ETH Zurich have integrated polymer strips, populated with surface mounted devices (SMD) and conductive tracks, during the weaving process [18]. The polymer strips were used as 'yarns', but their insertion restricts the shear behavior of the final textile fabric; and the use of a standard bare die (not encapsulated) limits the degree of bending that the strips can withstand. The electronics were exposed on the surface of the textile and rapidly failed after washing.

The maintenance of 'fiber-based' characteristics within an electronic textile is desirable [19]. Electronic yarn, (E-yarn) [20] was a further progression from the second generation of E-textiles that fulfilled this aim. Previously, conductive components and electronics replaced much or all of the textile fibers within textile structures, altering the properties of the completed textile. The difference between this third generation product and previous generations of electronic textiles was that the electronics were included within textile yarns that retained their textile properties, unlike, say, fiber optics included within electronic textiles. Electronics were contained within small micro-pods within E-yarn, leaving more than 90% of the volume of the E-yarn as textile fibers. The textile properties of E-yarn meant that it could be processed in knitting and weaving machines. The advantage of creating a textile-based E-yarn was that yarn properties could be retained to a greater extent than when using other materials. In particular:

1. The mechanical properties of the textile yarn were not adversely affected by the inclusion of electronics.
2. Moisture wicking and moisture retention occurred as in a normal textile yarn.
3. The E-yarn could be colored using textile dyeing methodology.

This allowed for electronics to be integrated in a highly discrete way, with electronic yarns being undetectable to the end user. Several types of E-yarn have been incorporated into a variety of textiles, as reported elsewhere in the literature.

E-yarn Structure and Manufacturing Process

Figure 1a shows a schematic of the E-yarn structure, with a photo of an E-yarn containing an LED in Figure 1b.

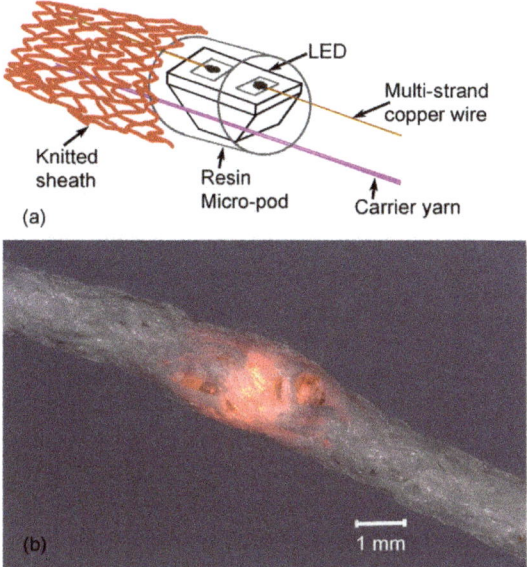

Figure 1. (**a**) A schematic showing the E-yarn (electronic yarn) structure, with an LED (light-emitting diode) protected by a micro pod and surrounded by a knitted sheath. (**b**) A completed E-yarn containing an LED (illuminated) shown at 30x magnification.

The production of E-yarn consisted of four main steps [21]: Firstly, a package die was soldered onto a fine copper wire. Encapsulation of the die and solder joints followed. The final two stages involved the addition of textile yarns twisted around the copper wire and attached micro-pods, before insertion of the construction into a soft, fibrous sleeve using a small-diameter circular warp knitting machine (RIUS; Barcelona, Spain). This gave the resulting E-yarn a textile feel, so that it could then be used in knitting and weaving machinery for garment production. The flow chart in Figure 2 shows the main steps in the process. The twisting of textile yarns around the core was not required for all E-yarn constructions, so it is shown within a box surrounded by dashed lines in Figure 2.

This work concentrates on the production of E-yarns. The E-yarns included within some of the prototypes in the literature were fabricated using a laborious hand-crafting process; this is both time consuming and may induce issues of repeatability into the production process. By automating the production process, a larger quantity of yarns can be produced, which will be required for industrial adoption. This has been ongoing work over many years, and as a result, an overview of the automation process is available elsewhere in the literature [21], as well as details of some prototypes produced with early versions of the automated process [22].

This paper will specifically concentrate on the automation of the encapsulation process. Here, encapsulation refers to the process of applying a fluid sealing compound into a small and pre-defined area around a die to produce a micro-pod. This encapsulated the die and the solder joints that attached it to copper wire. The sealing compound protected the electrical component from environmental effects, such as humidity and dust, and meant that the resulting E-yarn could be washed. Other benefits included improved electrical insulation, reliability, and protection against damage. The encapsulation was required to cover dies and solder joints only, leaving the wire interconnections between dies free to flex within the E-yarn. The resulting E-yarn consisted of more than 95% fibers, as the encapsulation volume was minimal.

The micro-pod was also used to create a bond between the copper wire and a carrier yarn that carried the mechanical stress in the E-yarn. The carrier yarn was placed alongside the copper wire, so that it became included within the micro-pod during the encapsulation process. The purpose of the encapsulation was to enhance the robustness and reliability of the E-yarn in the final product.

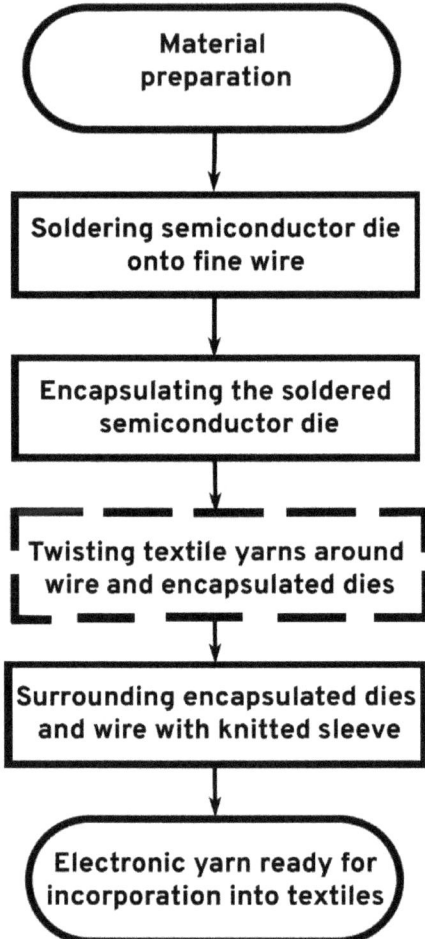

Figure 2. The main processes in the production of E-yarn. The dashed lines represent an optional stage: Twisting of textile yarns around the wire and encapsulated dies.

The aim of the work was to automate the stage in the E-yarn production process in which the package dies soldered to copper wire were covered in resin micro-pods. The resulting process allowed for repeated, automated production of micro-pods, increasing the E-yarn production speed from the previous manual process and paving the way for continuous, reliable, and repetitive production of E-yarn. This paper outlines key design considerations.

Ultimately, an automated process was created to form micro-pods around package dies that had been previously soldered to copper wire. The dies were encapsulated within a cylindrical mold, then removed using this automated process. A carrier yarn included within the encapsulation was shown to increase the tensile strength of the construction, enabling removal of the micro-pod from the mold. FTIR (Fourier Transform Infrared) analysis was used to assess optimal curing times for the resin micro-pods.

2. Design Considerations and Methodology

A methodology was developed to automate the encapsulation of semiconductor dies soldered onto multi-strand copper wire. This required an approach in which materials and machinery were selected, in conjunction with decisions about optimal processes to form the encapsulation micro-pods, and the order of operations. The challenge was to demonstrate that E-yarn production could be automated, so that the manufacturing process could ultimately be scaled up and commercialized. The selection of machinery and methods was based on the availability within the time and budgetary constraints of this project.

The core materials of E-yarn were copper wire and semiconductor dies. A 7-strand copper wire with a 50 µm strand diameter (Knight Wire, Potters Bar, UK) was chosen due to its flexibility compared with a single-strand wire of the same diameter. Semiconductor package dies of size 0402 were used. The examples shown in this work are a Kingbright KPHHS-1005SURCK Red LED, 630 nm 1005 (0402), Rectangle Lens package (RS Components, Corby, UK), and a thermistor (Murata 10 kΩ 100 Mw 0402 SMD NTC thermistor; part number NCP15XH103F03RC; Murata, Kyoto, Japan). This wire and these semi-conductor devices were based on those previously chosen in another work [21].

2.1. Resin Selection

The selection of the encapsulation material involved consideration of:

- Ability to dispense small amounts of encapsulant around the package dies;
- A method of curing that could be carried out without damage to the package dies;
- The speed of the curing process;
- Fabrication of a micro-pod that could withstand washing;
- Creation of a micro-pod that could transmit light from a package LED;
- Flexibility, for compatibility with surrounding textile materials.

The choice of the curing method for the encapsulation material was considered first. Encapsulating dies using a heat-curing method was undesirable due to the negative effect of the heating process on the die itself, and the time taken for many heat-curable resins to cure fully. Heat curing could also have led to mechanical failure of electronic components due to temperature-induced elastic or plastic deformation [23]. For these reasons, a photo-initiated curing method was preferred. Additional advantages included a low operational cost, easy maintenance, and the small footprint of space required for machinery. Moreover, an ultra-violet (UV)-curable coating reduced solvent emission, since most of the formulation was composed of oligomers and reactive diluents [24].

A UV-curable polyurethane acrylate system is composed of three basic components [24,25]:

- A resin, such as an oligomer or prepolymer, which is an unsaturated double bond or cyclic structure capable of ring opening;
- Reactive diluents (monomers with varying degrees of unsaturation), which have two functions: one is to reduce the viscosity of the systems, while the other function of the diluents is render crosslinking;
- A photo initiator, which absorbs UV radiation and generates reactive species that can initiate the polymerization.

Commercially-available UV-curable resins can be categorized by the polymerization mechanisms through which curing takes place. One uses radical polymerisation of monomers, such as acrylates or unsaturated polyesters. The other main method uses cationic polymerisation of multifunctional groups, such as epoxides and vinyl ethers [26]. An acrylated urethane was chosen for this research: Dymax 9001-E-V3.5 (Dymax, Torrington, UK). This is transparent to visible light, making it suitable for encapsulation of LEDs. Moreover, this resin is flexible, so it is suitable for textile applications. The resin is also used in bonding applications in the electronics industry, meaning that it has been fully tested on a range of electronic devices [27].

2.2. Curing the Resin

The chosen Dymax 9001 resin was designed to be cured with light at wavelengths from 320–450 nm [27]. Curing was carried out in the center of this range, with a 385 nm UV (BlueWave® QX4® LED Multi-Head Spot-Curing System; Intertronics, Kidlington, UK). The time required to cure the resin fully was assessed initially by checking that liquid resin became solid after the curing process. An investigation of the optimal curing time was carried out using Fourier Transform Infrared Spectroscopy (FTIR). For these experiments, 3 µL of resin (typical for a 1 mm diameter cylindrical encapsulation) was dispensed onto a plate of PTFE (polytetrafluoroethylene). Samples were cured for 10 s, 20 s, 30 s, and 50 s by exposure to the 385 nm UV light source. The samples were then kept in black bags to avoid any curing by visible light, before their transmission spectra were analyzed in an FTIR spectrometer (Perkin Elmer Spectrum Two; Llantrisant, UK).

2.3. Encapsulation Mold Design

Resin micro-pods can be formed by applying a small amount of resin around a die or by placing the die (soldered to copper wire) within a mold, then injecting resin into the mold. The latter method was chosen to ensure uniformity of the micro-pod size. The design of this encapsulation mold was a crucial part of the process, with careful consideration paid to the shape, size, and orientation. A cylindrical micro-pod shape was chosen for encapsulating the soldered semiconductor, as this shape had minimal edges. This shape was also easier to pass through the small-diameter warp-knitting machine that was used in a later stage of production of E-yarn, when the micro-pods were surrounded by a knitted sleeve. Cylindrical micro-pods were also more likely to be comfortable to the end users as the circular cross-section of the encapsulation was compatible with the cross-section of traditional textile yarn. The diameter of the encapsulation had to be minimized to keep the overall diameter of the yarn as thin as possible, thus suitable for embedding within ordinary garments: A thicker encapsulation would lead to a thicker final yarn. Conversely, the diameter of the mold had to allow the resin to flow in a way that would fully cover and protect the electronic component.

2.3.1. Mold Material

The selected cylindrical micro-pod shape could be formed within a tubular mold. A mold material was required that minimized adhesion of the cured resin to the mold wall to minimize the force required to release the micro-pod from the mold. The wetting properties of the inner mold surfaces were important for the resin dispensing process. The mold walls were also required to be UV-transparent to permit curing of the resin. Two potential mold materials were examined that fulfilled these criteria, and were available in tube form (which was the preferred shape for the mold; discussed in 2.3.2, below): Silicone (2mm Silicone Tube, part number a16090800ux0404; Sourcingmap, Mountain View, CL, USA) and PTFE (RS PRO Long Coil Tubing Without Connector, Fluoropolymer 22 bar, $-40 \rightarrow +150$ °C; RS Pro, Corby, Northants, UK). The internal surface structures of both materials were assessed using a scanning electron microscope (Scanning Electron Microscope, JEOL JSM-840A, SEMTech Solutions, Billerica, MA, USA)

2.3.2. Mounting of the Mold

The flexible tube that was chosen for the mold required mounting within a stable structure through which UV light could penetrate. The tube was placed within the body of a polypropylene syringe (Metcal PP Adhesive Dispenser Syringe; Techcon Systems, Garden Grove, CL, USA), with the back of the syringe removed to facilitate later mounting within an automated encapsulation system. The tube was held in place within the syringe using Transil 40-1 silicone elastomer (Mouldlife, Suffolk, UK). Transil 40-1 is a two-component silicone elastomer that crosslinks at room temperature by a poly-addition reaction to form a transparent, elastic material. The transparency was an advantage, minimizing UV blocking during the curing process, which was critical for this application. Transil

40-1 A and Transil 40-1 B were mixed by weight in a fixed ratio of 1:10 of parts A and B, respectively. The mixture was then degassed for 10 min and poured slowly into the outer part of a plastic syringe, with the flexible tube held inside the syringe. Figure 5 shows the silicone mold around the flexible tube in which encapsulation was carried out. The whole syringe, with the silicone and the flexible tube inside it, were mounted vertically within a bracket. Vertical mounting ensured that the resin flowed evenly over the die, rather than falling to one side. An image of the mold mounted to the encapsulation system is shown later in this work (Section 3).

2.4. Resin Dispensing

There was a need to inject a small, pre-determined volume of resin into the mold to surround each package die held within it. A preeflow® eco-PEN Precision Volumetric Dosing Pump (Intertronics, Kidlington, Oxfordshire, UK) was used to apply resin and create a micro-pod around the electronic components. This system used volumetric, positive-displacement dosing, which gave both accuracy and repeatability. This dosing system could dispense small quantities, with a minimum volume of 0.001 mL, with flow speeds of 0.12–1.48 mL/min.

2.5. Carrier Yarn

To enhance the mechanical properties of the completed E-yarn, the soldered semiconductor die was encapsulated along with a carrier yarn. The role of the carrier yarn was to support the soldered joint, protecting it and the copper wire from mechanical stresses during demolding and during subsequent E-yarn formation processes. 100 Dernier Vectran™ (Kururay, Tokyo, Japan) was chosen for its high tensile strength [21] combined with its flexibility. Figure 3 shows a soldered semiconductor die (a thermistor) and Vectran™ carrier yarn running alongside the copper wire; both included within a resin micro-pod that forms the encapsulation.

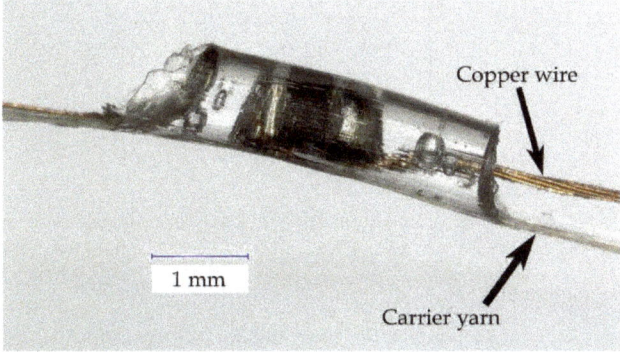

Figure 3. A carrier yarn included within the encapsulation of a semiconductor soldered onto copper wire, at 50× magnification.

2.6. Tensile Force Required to Remove the Micro-pod from the Mold

Pulling on the carrier yarn provided a method of removing encapsulated dies from the mold in which the micro-pod was formed. Experiments were carried out to find the tensile forces required to remove cured micro-pods from the mold. The results were compared with the tensile forces required to break the carrier yarn. This established the limitations of the molding system; indicating at which micro-pod length friction forces would prevent removal from the mold using the carrier yarn to provide the demolding force. To find the tensile force required to demold the micro-pod, the mold was mounted in a zwickiLine tensile tester (Z2.5; Zwick/Roell, Ulm, Germany). Micro-pods were pulled from a mold as shown in Figure 4a. Measurements were taken of the tensile force required to remove encapsulated dies from the mold. This force was expected to increase in proportion to the amount of contact surface between the cured resin and the mold tube walls. Figure 4b shows this contact area.

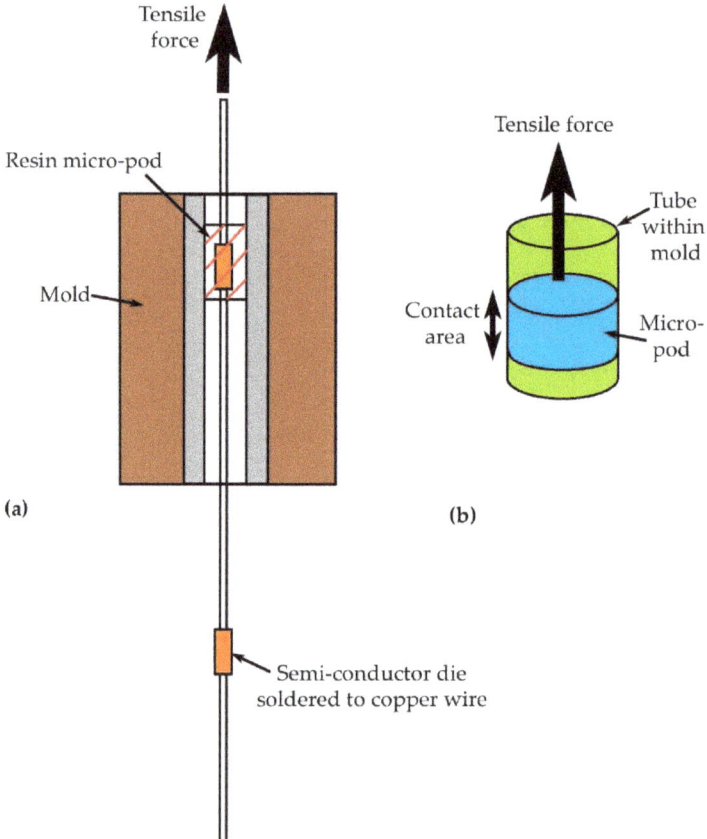

Figure 4. (a) A schematic showing an encapsulated die being pulled from the mold. (b) Diagram showing the area over which friction forces will act when pulling a die from the mold.

2.7. Automating the Production Process

To show the feasibility of automating the encapsulation process, which is critical for commercial production, the encapsulation system was partially automated. A prototype automated encapsulation unit was designed in which the electronic components to be encapsulated were moved into a mold, a pre-determined quantity of resin was dispensed, the resin was UV-cured, and the micro-pod was removed from the mold (completing the encapsulation process). The design illustrated in this paper is based on experimental work in which several designs were investigated. The focus is placed on the design that gave the most consistent results. A schematic of the automated process is shown in Figure 5.

To allow for automation, the location of the soldered die needed to be determined to ensure that resin would only be dispensed once the die was in the correct location. A transmissive fiber sensor unit, FU-58, was used with an amplifier, FS-N11MN (both from Keyence; Milton Keyence, UK). This fiber optic sensor detected changes in light intensity. When the light intensity dropped due to a die arriving between the two parts of the fiber optic sensor, the voltage dropped, so a signal was sent to the encapsulation unit. The signal was interpreted using a LabView program (Version 14; National

Instruments, Newbury, UK). The sensor detected the soldered die inside the transparent mold tube, at the point at which encapsulation took place. The minimum detectable object was 0.005 mm^2, which was adequate for all package die sizes of interest at this time.

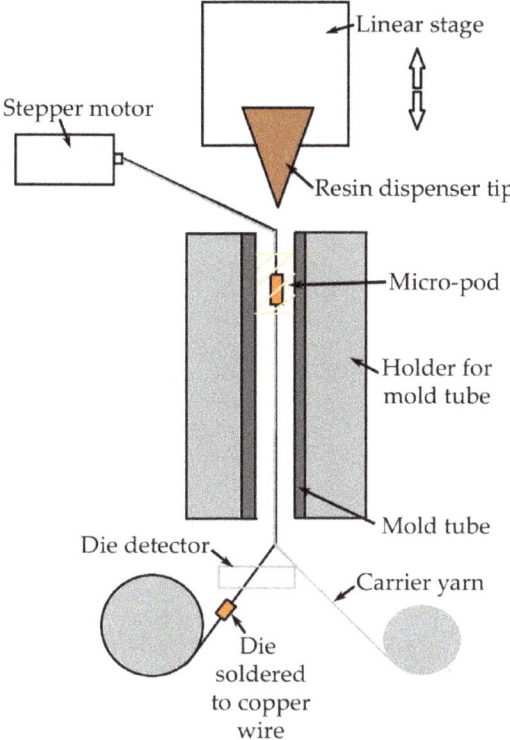

Figure 5. Schematic illustration of the automated encapsulation unit, showing die detection before entering the silicone mold; and a completed micro-pod surrounding an encapsulated die about to exit from the mold.

Computer controlled elements were used to automate the encapsulation process. The resin dispenser unit was mounted onto a motorized linear translation stage with an integrated controller (LTS 15; Thorlabs, Ely, UK). This enabled the resin dispenser to be moved vertically to a specific encapsulation point. The linear translation stage provided 150 mm of linear travel with a maximum vertical velocity of 3 mms^{-1}. The force applied by the stepper motor ensured that the linear translation stage remained fixed when no power was supplied, so that a brake did not need to be employed to keep the stage in position. This would not have been the case if a DC (direct current) servo motor was used. The linear translation stage featured an integrated electronic controller that could be controlled using a PC or manually. The stage was calibrated by the manufacturer, giving a typical error of ±4.0 μm when moving over ±20.0 μm travel.

A second set of motors was used to move the copper wire (with attached, soldered electronic components). A stepper motor (ISM 7411E; National Instruments Corporation (U.K.) Ltd., Newbury, UK) was chosen to drive the soldered component to a specific point where the encapsulation process could take place. This stepper drive motor had micro-step emulation and an integrated encoder, and was chosen as it could give a feedback signal to the driving software and therefore increase the accuracy of the movement. An Ethernet communication port controlled the motor via LabVIEW.

Process Control

The automated encapsulation process was controlled and monitored by a LabView program, which controlled the sequence task as illustrated in the flow chart in Figure 6. The process consisted of the following steps: Following detection of the die, once the electronic component arrived at the encapsulation point in the mold, the stepper motor stopped and the linear stage moved the dispenser until the dispenser tip reached the encapsulation point where the soldered die was located. Resin was then dispensed. The linear stage was used to move the dispenser up again to avoid curing the resin within the dispenser's tip. The LED UV light source was used to cure the dispensed resin in the mold. The process was then repeated to encapsulate the next die.

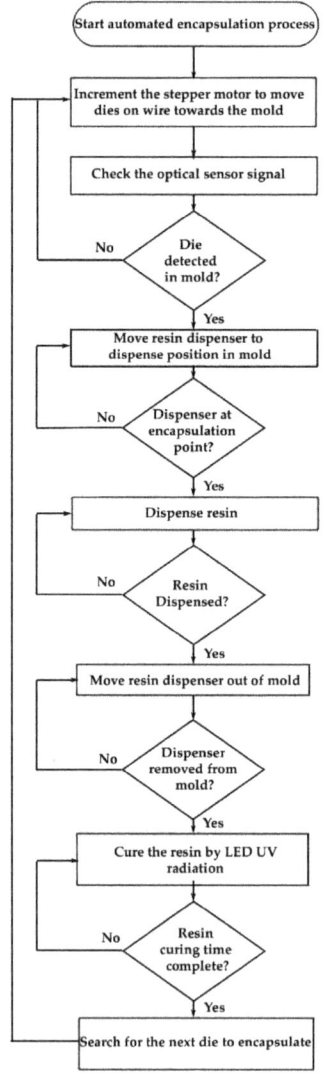

Figure 6. Flowchart illustrating the sequence of logic steps followed by the LabView program to enable the automated encapsulation of components.

2.8. Tensile Testing

The solder joints created in the first stage of production were expected to be the weakest point in the ensemble. The encapsulation process was expected to substantially increase the soldered joint strength against mechanical stresses. Tensile tests were carried out to establish the changes in tensile strength of the construction at each stage of production. The zwickiLine tensile tester was used, working to standard ASTM E8 [28], which related to 'Standard Test Methods for Tension Testing of Metallic Materials'. The test speed was reduced to 50 mm min^{-1} to enable testing of the solder joints included within the E-yarn. Five samples were measured from each stage of the E-yarn production process:

- Bare semiconductor dies soldered to copper wire;
- Encapsulated dies, with carrier yarn included within the encapsulation;
- Completed E-yarn, with a textile outer covering around an encapsulated die.

The tensile forces required to break each sample were compared.

3. Results

A method was developed to encapsulate package dies on flexible wire, with the process automated using the machinery shown in Figure 7. This shows copper wire and carrier yarn being fed through a sensor and into the bottom of the mold, in which the encapsulation is carried out. The system for injecting resin is directly above the mold, and the UV light guide points to the mold. It was found that a dispensed micro-dot of resin could flow into the tube that formed the mold, rather than being retained where the dispenser injected it. The requirement for resin to flow into the tube, but not fall straight through, also placed a maximum limitation on the diameter of the mold, which was required to hold the liquid resin inside until curing was complete. The chosen electronic components, such as package LEDs and thermistors, were small enough (1.0 × 0.5 × 0.5 mm) to be encapsulated within a 1.27 mm internal-diameter mold.

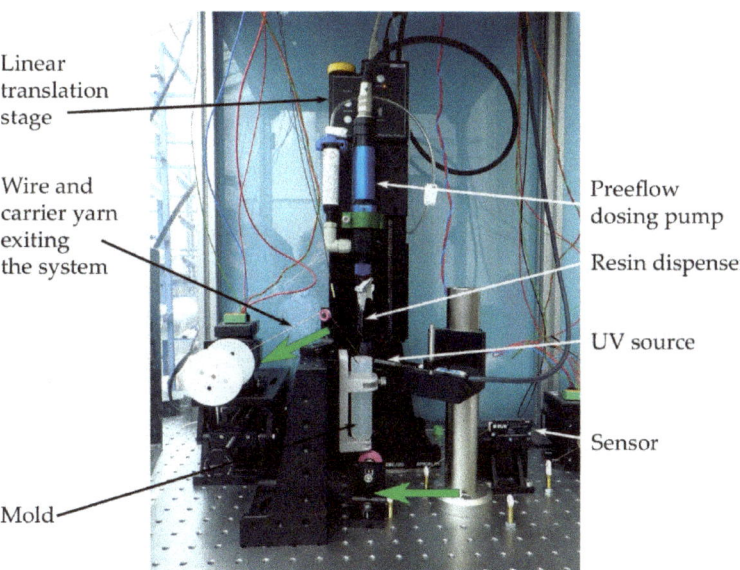

Figure 7. A photograph of the encapsulation system. The main components are labelled, and the large, green arrows show the direction of travel through the system of the copper wire with attached sensors.

The method of detecting the presence of a package die to initiate the encapsulation process was achieved with an accuracy of 98% for thermistors and 87% for LEDs. The sensitivity and accuracy of the process depended on the color, reflectivity, size, and shape of the component. The transparent lenses of the LEDs made detection by changes in light intensity more difficult. The angle at which the die arrived at the sensor also affected the detection ability. The front or back of the die facing the sensor was likely to lead to detection, whilst the side of a die was less likely to be detected.

3.1. Fourier Transform Infra-Red (FTIR) Analysis of UV Curing Times

Fourier transform infra-red (FTIR) analysis was performed on sample micro-pods that had been cured for 10 s, 20 s, 30 s, and 50 s by exposure to the 385 nm UV light source. Figure 8 shows transmission spectra for the samples cured for 10 s and 50 s. The curves for the samples cured for 20 s and 30 s were very similar to that for the sample cured for 10 s, so are omitted from Figure 8 for clarity. The peaks of all the analyzed samples are similar to spectra found in the literature for acrylated urethane resins [29]. The spectra in Figure 8 are all similar, implying that full curing of the resin had occurred in all cases. It was therefore acceptable to use a 10 s curing time for a sample of this size, and possibly shorter exposure times may also be appropriate. It was interesting to note that for 50 s of curing (the green line on the graph), there was some reduction in the trough at 1722 cm^{-1}. This indicated over-curing as the literature states that the number of double bonds (C=C) at 1635 cm^{-1} and 810 cm^{-1} decrease when over-curing occurs [29]. Over-curing of the resin can affect the structural and optical properties of the resin. Over-curing is to be avoided, as this may change the mechanical properties and optical characteristics of the resin, for example, by causing yellowing. Minimizing the UV curing time for encapsulation had two advantages: Increased speed of encapsulation and prevention of UV degradation of the micro-pods.

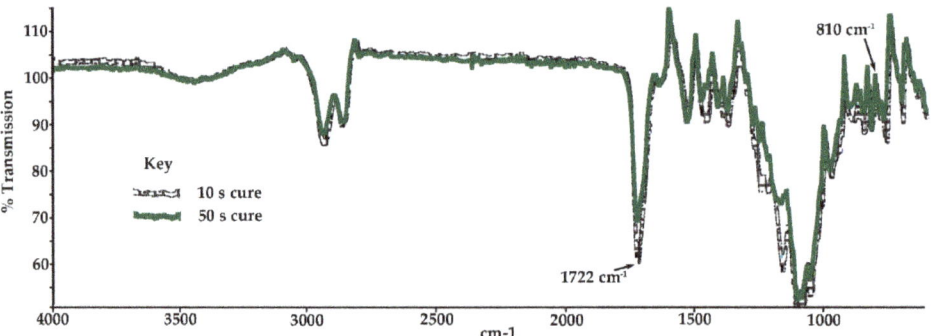

Figure 8. FTIR results from resin samples cured for 10 and 50 s.

3.2. Force Required to Extract Micro-pods from the Mold

Two mold materials were explored in this work (silicone and PTFE), with the internal surfaces of both materials assessed using a scanning electron microscope. Figure 9 shows the internal surfaces of the tubing made from (a) PTFE and (b) silicone. There are clear structural differences between the two surfaces, with the surface of the PTFE tube appearing much rougher than that of silicone at 10,000 times magnification. It was undesirable for the resin to stick within the mold, as this would increase the tension required to remove the micro-pod after curing, increasing the likelihood of a breakage occurring. Both mold materials were tested for ease of removal of a micro-pod from a mold. It was found that the carrier yarn stuck to the wall of the silicone tube during molding, rather than staying within the encapsulant. It was possible that the rougher surface of the PTFE could provide less points of contact area for the cured resin micro-pod to contact, especially if the resin shrank on curing. PTFE was therefore chosen as the mold material, despite its rougher appearance.

Figure 9. SEM images of the inner surface of the encapsulation mold materials at 10,000 times magnification: (**a**) PTFE (polytetrafluoroethylene) tube. (**b**) Silicone tube.

The force required to pull micro-pods from the PTFE mold was subsequently assessed using tensile testing. Figure 10 shows the force required to extract micro-pods from the mold after curing the resin inside a 1.27 mm PTFE tube. The increasing values of tensile force in Figure 10 shows that increasing volumes of resin, which led to increased micro-pod lengths (from 3.9 to 15.8 mm), and therefore contact area, required an increased force to be applied to remove the cured resin from the PTFE tube. The tensile force should not exceed the breaking force of the carrier yarn, which was known to be between 26.2 and 27.6 N [21]. The graph in Figure 10 shows that all micro-pods under investigation could be removed from the mold tube by exerting forces below these values, but further increases in the volume of resin or the diameter of the tube could lead to the force required to remove the micro-pod from the mold exceeding the carrier yarn strength. The results also indicated that the main force required to remove the micro-pod from the tube was a static force: The force required to initiate motion [30]. Little kinetic force [30] was required to maintain motion after this.

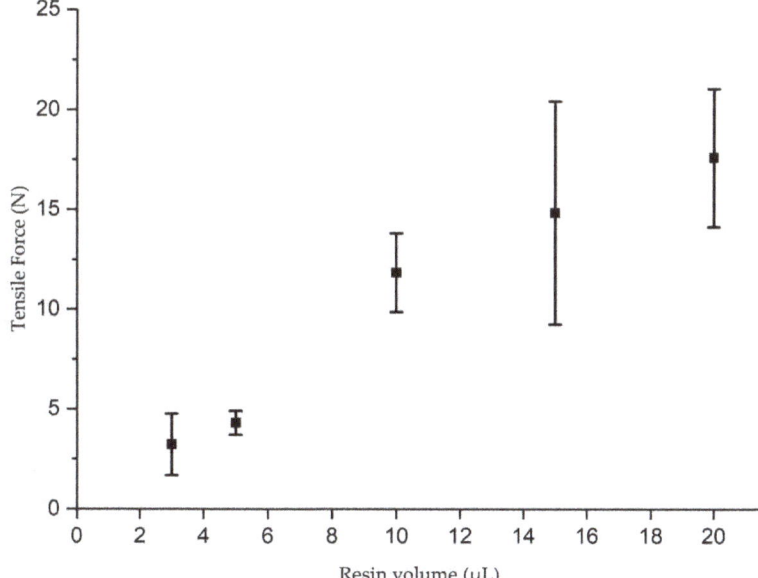

Figure 10. Tensile forces required to pull a micro-pod from a mold as a function of the quantity of resin used to encapsulate a semiconductor in a 1.27 mm-diameter PTFE tube.

3.3. Tensile Testing of E-yarn Components

As a final validation of the influence of the encapsulation process, the tensile properties of the encapsulated yarn were compared with those of soldered, un-encapsulated components. Tensile tests were performed on five samples of un-encapsulated dies; five encapsulated dies; as well as five completed E-yarns containing encapsulated dies with carrier yarn, within knitted sleeves. The results are shown in Figure 11. The solid lines at the base of the graph show that un-encapsulated dies soldered onto wire had a low tensile strength with maxima between 2.72 and 3.21 N. Two of the five specimens broke after elongation of the copper wire rather than by fracture of the solder joints between the die and the wire. This is shown by the solid lines that extend along the x-axis of the graph, as the ductile copper extended before breaking. Encapsulating the dies within micro-pods increased the tensile strength to 15.01 to 23.94 N, as shown by the dashed lines on the graph. Encapsulation of the soldered joints, as well as carrier yarns included within the micro-pods, increased the tensile strength of these specimens. The completed E-yarn had a much greater tensile strength, as indicated by the dotted lines on the graph, showing initial breaking forces from 54.93–67.46 N. The use of both the carrier yarn and an outer, knitted sleeve greatly increased the breaking strength of the construction. The E-yarn sleeve was manufactured from eight individual polyester yarns plus packing yarns. The jagged pattern of breakage is likely due to fiber-fiber friction between the packing yarns within the construction. The knitting process brought these yarns together, creating a structure with greater tensile strength than that of the individual textile yarns that made up the construction. The single strands of copper and Vectran™ at the core of the E-yarn had a lower tensile strength. (Vectran™ yarn has been shown previously to have a tensile strength of 23–28 N [21].) The function of the micro-pod was not to provide additional tensile strength, but to protect the die and solder joints against impact and moisture ingress, so the outer sleeve was important in increasing the overall tensile strength.

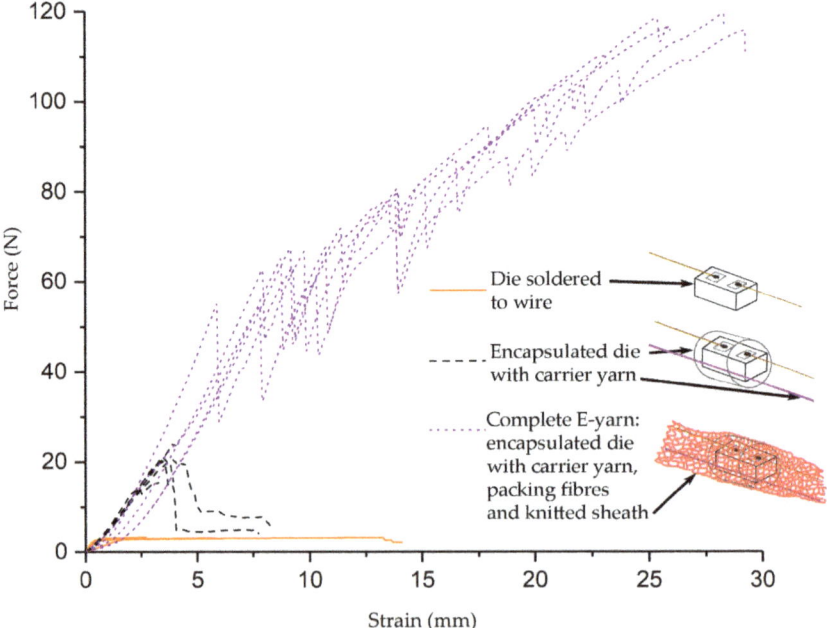

Figure 11. Tensile test results for samples of soldered dies; encapsulated dies with carrier yarn; and E-yarn with a knitted sheath surrounding encapsulated dies with carrier yarn and packing fibers.

4. Discussion

Ultimately, an encapsulation unit was designed and constructed, suitable for automated encapsulation of semiconductor dies that had been soldered to wire. This enabled fast, repeated encapsulation of dies soldered to copper wire. The design of the mold in which the micro-pod was formed was key to this process. The design was optimized for 0402 metric package dies measuring 0.5 × 0.5 × 1.0 mm, for which a 1.27 mm internal diameter tube was used as a mold. Since the diameter of the tube was fixed, the volume of resin dispensed, in addition to the size of the component, determined the finished length of the encapsulation. One method of incorporating complex electronic circuits into a yarn is to use flexible Kapton strips as a substrate (as recently reported in the literature [31,32]). This would require micro-pods with lengths from 10 mm up to at least 50 mm. The removal of longer micro-pods from the mold requires tensile forces that exceed the strength of the carrier yarn, so further development of the methodology is required. Further work is also required to investigate the effects of altered mold sizes for smaller and larger dies. The use of a thinner tube to encapsulate dies smaller than the 0402 metric package size used in this work may affect the flow of resin. Larger dies, requiring greater tube diameters, may require alterations to the design to prevent resin from flowing out of the tube before curing.

The prototype automated encapsulation unit described in this work offered a substantial improvement over the manual process previously used to fabricate E-yarns, substantially increasing the speed of the process. This technology is suitable for encapsulation of small electronic dies, such as 0402 metric package dies. This will help lead the way towards larger volume production of E-yarn, which is a critical step towards future commercialization and a greater uptake of the technology.

The E-yarns produced using the encapsulation process were designed to be incorporated into fabric using knitting and weaving machinery. The feasibility of this was demonstrated by inserting E-yarns containing thermistors into knitted textile thermographs [33], and by weaving, using E-yarns containing photodiodes in the weft of woven structures [34]. An important function of the micro-pods formed using the encapsulation unit was to protect the package dies from moisture ingress. Wash trials involving machine washing and tumble drying were carried out on five LED-yarns [35], and on photodiode yarns [34]. Both tests demonstrated that micro-pods formed during the encapsulation process can protect the enclosed package dies from moisture ingress that would lead to failure. Failures were observed at the interface between the copper wire and the micro-pod [34].

5. Conclusions

The manufacture of E-yarn requires the creation of resin micro-pods. These surround package dies that have been soldered to copper wire, protecting the die and solder joints against abrasion and moisture ingress. A prototype, automated encapsulation unit was developed. This included a mold with an inner PTFE tube, in which micro-pods were formed and underwent UV-curing. The materials and construction of the mold were chosen to enable formation, curing, and removal of the resin micro-pod. A Vectran™ carrier yarn with a high tensile strength was placed alongside the copper wire, with a portion of the yarn included within the micro-pod. This added tensile strength to the construction, including when pulling the micro-pod out of the mold. Experiments showed that the force required to pull the micro-pod from the mold was proportional to the volume of resin used to make the micro-pod, with 15 N required to pull a 15 µL micro-pod from a 1.27 mm-diameter mold. FTIR analysis was used to assess whether the resin within the micro-pod was completely cured. This showed that a 10 s cure time was adequate for a cylindrical micro-pod made in a 1.27 mm-diameter mold. Automation of the encapsulation process enabled repeated encapsulation of dies previously soldered to copper wire. Incoming dies were detected 87% to 98% of the time. This process improved the ability to produce the amounts of E-yarn required for further creation of prototype electronic textiles.

Author Contributions: T.D. and T.H.-R. supervised the program of study and provided specialist technical knowledge essential for the completion of this work. M.-N.N. and D.A.H. designed the study, conducted the experiments and performed the data analysis. M.-N.N., T.H.-R. and D.A.H. wrote the paper. All authors discussed the results and contributed to producing the final manuscript.

Funding: This research was funded by the Engineering and Physical Sciences Research Council (EPSRC) grant EP/M015149/1 Novel manufacturing methods for functional electronic textiles.

Acknowledgments: The authors would like to thank our colleagues from the Advanced Textiles Research Group (ATRG) at Nottingham Trent University, who provided insight and expertise that greatly assisted the research, especially Ioannis Anastasopoulos, Carlos Oliveira and Richard Arm. M.-N.N. thanks CARA for their support, sponsorship and financial support. The authors would also like to thank Kathryn Kroon for her assistance in acquiring the SEM images shown in Figure 9.

Conflicts of Interest: The authors declare no conflict of interest. The funders had no role in the design of the study; in the collection, analyses, or interpretation of data; in the writing of the manuscript, or in the decision to publish the results.

References

1. Park, S.; Jayaraman, S. Smart Textiles: Wearable electronic Systems. *MRS Bull.* **2003**, *28*, 585–591. [CrossRef]
2. Stoppa, M.; Chiolerio, A.; Stoppa, M.; Chiolerio, A. Wearable Electronics and Smart Textiles: A Critical Review. *Sensors* **2014**, *14*, 11957–11992. [CrossRef] [PubMed]
3. Fernández-Caramés, T.; Fraga-Lamas, P.; Fernández-Caramés, T.M.; Fraga-Lamas, P. Towards The Internet-of-Smart-Clothing: A Review on IoT Wearables and Garments for Creating Intelligent Connected E-Textiles. *Electronics* **2018**, *7*, 405. [CrossRef]
4. Fairs, M. Bono's Laser Stage Suit by Moritz Waldemeyer. Available online: https://www.dezeen.com/2010/02/28/bonos-laser-stage-suit-by-moritz-waldemeyer/ (accessed on 20 December 2018).
5. Castano, L.M.; Flatau, A.B. Smart fabric sensors and e-textile technologies: A review. *Smart Mater. Struct.* **2014**, *23*, 053001. [CrossRef]
6. Weng, W.; Chen, P.; He, S.; Sun, X.; Peng, H. Smart Electronic Textiles. *Angew. Chem. Int. Ed.* **2016**, *55*, 6140–6169. [CrossRef] [PubMed]
7. Hughes-Riley, T.; Dias, T.; Cork, C. A Historical Review of the Development of Electronic Textiles. *Fibers* **2018**, *6*, 34. [CrossRef]
8. Mi Pulse Smart Bras for Smart Fitness. Available online: https://www.mi-pulse.com/ (accessed on 17 January 2019).
9. The Liten Institute Printed Components. Available online: http://liten.cea.fr/cea-tech/liten/en/Pages/technoNanotechComponents/PrintedComponents.aspx (accessed on 17 January 2019).
10. Lind, A.H.N.; Mather, R.R.; Wilson, J.I.B. Input energy analysis of flexible solar cells on textile. *IET Renew. Power Gener.* **2015**, *9*, 514–519. [CrossRef]
11. Takamatsu, S.; Kobayashi, T.; Shibayama, N.; Miyake, K.; Itoh, T. Fabric pressure sensor array fabricated with die-coating and weaving techniques. *Sens. Actuators A Phys.* **2012**, *184*, 57–63. [CrossRef]
12. Wijesiriwardana, R.; Mitcham, K.; Dias, T. Fibre-Meshed Transducers Based Real Time Wearable Physiological Information Monitoring System. In Proceedings of the Eighth International Symposium on Wearable Computers, Arlington, VA, USA, 31 October–3 November 2004; pp. 40–47.
13. Gopalsamy, C.; Park, S.; Rajamanickam, R.; Jayaraman, S. The Wearable Motherboard?: The first generation of adaptive and responsive textile structures (ARTS) for medical applications. *Virtual Real.* **1999**, *4*, 152–168. [CrossRef]
14. Rein, M.; Favrod, V.D.; Hou, C.; Khudiyev, T.; Stolyarov, A.; Cox, J.; Chung, C.-C.; Chhav, C.; Ellis, M.; Joannopoulos, J.; et al. Diode fibres for fabric-based optical communications. *Nature* **2018**, *560*, 214–218. [CrossRef]
15. Bigham, K.J. *Drawn Fiber Polymers: Chemical and Mechanical Features*; Zeus Ind. Prod. Ltd.: Orangeburg, SC, USA, 2018.
16. European Commission. Integrating Platform for Advanced Smart Textile Applications. Available online: https://cordis.europa.eu/project/rcn/95473/factsheet/en (accessed on 16 January 2019).
17. Primo1D: The Technology. Available online: http://www.primo1d.com/e-thread/the-technology (accessed on 16 January 2019).
18. Zysset, C.; Kinkeldei, T.; Münzenrieder, N.; Petti, L.; Salvatore, G.; Tröster, G. Combining electronics on flexible plastic strips with textiles. *Text. Res. J.* **2013**, *83*, 1130–1142. [CrossRef]

19. Zeng, W.; Shu, L.; Li, Q.; Chen, S.; Wang, F.; Tao, X.-M. Fiber-Based Wearable Electronics: A Review of Materials, Fabrication, Devices, and Applications. *Adv. Mater.* **2014**, *26*, 5310–5336. [CrossRef] [PubMed]
20. Dias, T. Electronically Functional Yarns. Patent WO2016/038342 A1, 17 March 2016.
21. Hardy, D.A.; Anastasopoulos, I.; Nashed, M.-N.; Oliveira, C.; Hughes-Riley, T.; Komolafe, A.; Tudor, J.; Torah, R.; Beeby, S.; Dias, T. An Automated Process for Inclusion of Package Dies and Circuitry within a Textile Yarn. In Proceedings of the 2018 Symposium on Design, Test, Integration & Packaging of MEMS and MOEMS (DTIP), Roma, Italy, 22–25 May 2018.
22. Dias, T.; Hughes-Riley, T. Electronically Functional Yarns Transform Wearable Device Industry. *R&D Mag.* **2017**, *59*, 19–21.
23. Minges, M.L. *Electronic Materials Handbook: Packaging*; CRC Press: Boca Raton, FL, USA, 1989; Volume 1, ISBN 9780871702852.
24. Srivastava, A.; Agarwal, D.; Mistry, S.; Singh, J. UV curable polyurethane acrylate coatings for metal surfaces. *Pigment Resin Technol.* **2008**, *37*, 217–223. [CrossRef]
25. Decker, C. Contributed papers UV-radiation curing chemistry. *Pigment Resin Technol.* **2001**, *30*, 278–286. [CrossRef]
26. Boyd, R.H.; Phillips, P.J. *The Science of Polymer Molecules: An Introduction Concerning the Synthesis, Structure, and Properties of the Individual Molecules that Constitute Polymeric Materials*; Cambridge University Press: Cambridge, UK, 1996; ISBN 0521565081.
27. Dymax Corporation ELECTRONIC ASSEMBLY MATERIALS: 9001-E-V3.5 Product Data Sheet: Multi-Cure® 9001-E-V3.5 Resilient, Clear Encapsulant. Available online: https://www.dymax.com/images/pdf/pds/9001-e-v35.pdf (accessed on 15 November 2018).
28. ASTM ASTM E8/E8M-16A Standard Test Methods for Tension Testing of Metallic Materials. 2016. Available online: https://www.astm.org/Standards/E8.htm (accessed on 25 September 2018).
29. Kunwong, D.; Sumanochitraporn, N.; Kaewpirom, S. Curing behavior of a UV-curable coating based on urethane acrylate oligomer: The influence of reactive monomers. *Songklanakarin J. Sci. Technol.* **2011**, *33*, 201–207.
30. Israelachvili, J.N.; Chen, Y.-L.; Yoshizawa, H. Relationship between adhesion and friction forces. *J. Adhes. Sci. Technol.* **1994**, *8*, 1231–1249. [CrossRef]
31. Li, M.; Tudor, J.; Liu, J.; Komolafe, A.; Torah, R.; Beeby, S. The thickness and material optimization of flexible electronic packaging for functional electronic textile. In Proceedings of the 2018 Symposium on Design, Test, Integration & Packaging of MEMS and MOEMS (DTIP), Roma, Italy, 22–25 May 2018; pp. 1–6.
32. Li, M.; Tudor, J.; Torah, R.; Beeby, S. Stress Analysis and Optimization of a Flip Chip on Flex Electronic Packaging Method for Functional Electronic Textiles. *IEEE Trans. Compon. Packag. Manuf. Technol.* **2018**, *8*, 186–194. [CrossRef]
33. Lugoda, P.; Hughes-Riley, T.; Morris, R.; Dias, T. A Wearable Textile Thermograph. *Sensors* **2018**, *18*, 2369. [CrossRef]
34. Satharasinghe, A.; Hughes-Riley, T.; Dias, T. Photodiodes embedded within electronic textiles. *Sci. Rep.* **2018**, *8*, 16205. [CrossRef]
35. Hardy, D.; Moneta, A.; Sakalyte, V.; Connolly, L.; Shahidi, A.; Hughes-Riley, T. Engineering a Costume for Performance Using Illuminated LED-Yarns. *Fibers* **2018**, *6*, 35. [CrossRef]

© 2019 by the authors. Licensee MDPI, Basel, Switzerland. This article is an open access article distributed under the terms and conditions of the Creative Commons Attribution (CC BY) license (http://creativecommons.org/licenses/by/4.0/).

Article

Engineering a Costume for Performance Using Illuminated LED-Yarns

Dorothy A. Hardy *, Andrea Moneta, Viktorija Sakalyte [†], Lauren Connolly [‡], Arash Shahidi and Theodore Hughes-Riley

School of Art and Design, Nottingham Trent University, Shakespeare Street, Nottingham NG1 4FQ, UK; andrea.moneta@ntu.ac.uk (A.M.); viktorija.sakalyte@hud.ac.uk (V.S.); Lauren_Connolly18@hotmail.co.uk (L.C.); arash.shahidi@ntu.ac.uk (A.S.); theodore.hughesriley@ntu.ac.uk (T.H.-R.)
* Correspondence: dorothy.hardy@ntu.ac.uk; Tel.: +44-115-84-82709
† Current address: School of Art, Design and Architecture, University of Huddersfield, Queensgate, Huddersfield HD1 3DH, UK.
‡ Ms. Connolly is now a Freelance Theatre Designer.

Received: 30 April 2018; Accepted: 29 May 2018; Published: 1 June 2018

Abstract: A goal in the field of wearable technology is to blend electronics with textile fibers to create garments that drape and conform as normal, with additional functionality provided by the embedded electronics. This can be achieved with electronic yarns (E-yarns), in which electronics are integrated within the fibers of a yarn. A challenge is incorporating non-stretch E-yarns with stretch fabric that is desirable for some applications. To address this challenge, E-yarns containing LEDs were embroidered onto the stretch fabric of a unitard used as part of a carnival costume. A zig-zag pattern of attachment of E-yarns was developed. Tensile testing showed this pattern was successful in preventing breakages within the E-yarns. Use in performance demonstrated that a dancer was unimpeded by the presence of the E-yarns within the unitard, but also a weakness in the junctions between E-yarns was observed, requiring further design work and reinforcement. The level of visibility of the chosen red LEDs within black E-yarns was low. The project demonstrated the feasibility of using E-yarns with stretch fabrics. This will be particularly useful in applications where E-yarns containing sensors are required in close contact with skin to provide meaningful on-body readings, without impeding the wearer.

Keywords: electronic yarn; E-yarn; LED-yarn; LED; stretch fabric; illuminated textiles; electronic textiles; E-textiles

1. Introduction

Incorporation of electronics within textiles offers an opportunity to integrate lighting into clothing and to monitor health through sensors embedded within clothing. The growing market for electronic textiles (E-textiles) demonstrates the high level of interest in this areas [1]. Thus, the need exists for further development of nonintrusive electronics that do not impede the wearer. Ideally, clothing incorporating electronics would have normal drape, conformability, and stretch. This is a challenge due to the considerable difference in material properties between textile fibers and electronics. One solution is to use electronic yarn (E-yarn) [2], which omits the circuit board substrate that forms a part of many electronic circuits. Instead, package dyes are attached onto a flexible copper wire to create what is effectively an electronic fiber, which is then contained within textile fibers. This creates an E-yarn with a textile feel and drape similar to that of other fibers within garments [3]. Development of a semi-automated process has led to the ability to produce E-yarns relatively easily and quickly [4]. Previously, E-yarns had been produced by a craft process [5], with six or fewer being used in most garments [6]. There was a need to demonstrate the feasibility of attaching tens of E-yarns onto one garment, illustrating how electronics could become a significant and integral part of clothing. Finding a method of attaching E-yarns to stretch fabric is also important. E-yarns are non-stretch due to the central, non-stretch copper wire, but integration with stretch fabrics would enable the use of sensors within E-yarns to take measurements of the human body from as close to the skin surface as possible. This would be especially useful in obtaining meaningful readings from sensors included within E-yarns by increasing the number and type of garments into which this technology could usefully be incorporated.

E-yarns are designed to be unobtrusive within textiles. In the project described in this paper, we investigated the use of the E-yarns in stretch fabrics and tested their functionality. This was achieved through use of E-yarns containing LEDs (LED-yarns), as shown in Figure 1. The illumination of the LEDs enabled the quick assessment of the functionality of the E-yarns. The chosen project was the design of an illuminated carnival costume for use in a competition to be held in a theatre [7]. This provided a time-delimited project for which funding was available, so that a garment containing multiple LED-yarns could be designed and made. This provided an ideal platform for testing methods of attaching LED-yarns onto stretch fabric that was placed next to the skin of a dancer wearing the costume. This provided a rigorous test of the durability of the E-yarns and their connections in a relevant operational environment, so that any subsequent recommendations for the use of E-yarn on stretch fabric could be considered with this as a benchmark. Displaying the costume in carnivals and other performances could achieve an additional aim of ensuring that the E-yarns were viewed by a large audience, increasing the visibility of this technology.

Figure 1. Illuminated light emitting diode (LED)-yarn shown next to a 30 mm long pin.

The initial brief for the project required the use of illuminated E-yarns, other methods for incorporating E-textile lighting exist and have been gradually developed [8,9]. The earliest example is the lighting worn by dancers in the ballet La Farandole in 1884 [10], showing use of filament lightbulbs in costume lighting applications. Electronic textiles for lighting typically use one of four main methods: fiber optics [11,12], LED strips [13], electroluminescent wires [14], and lasers [15]. Electroluminescent wires, fiber optics, and the use of lasers generate sufficient light to be clearly visible during theatre or outdoor night-time performances, but also restrict the flexibility and conformability of a garment. Similarly, the direct attachment of LEDs onto a costume using strips of LEDs or sewing individual LEDs into holders is non-ideal as it affects the textile material properties. Examples of costumes incorporating these elements include Bono's "laser suit" [16], and the Slovakian Tron Dance that incorporate LED-covered suits [13]. Fiber optics have been included within costumes that can accommodate the limited flexibility of the fiber optic elements, such as the Scottish Opera's Queen of the Night costume [12]. The lack of flexibility and conformity has limited the use of E-textiles within performance. Wider adoption of E-textiles has also likely been limited in part by cost and by the infancy of design philosophies and practices when using E-textiles [17].

Smaller and cheaper microelectronics have allowed some electronic garments and wearable accessories to enter the marketplace, with aesthetics being the primary function. These have principally been featured in the clubbing scene, with an example being the Sound Activated T-shirt [18]. High-end alternatives also exist, such as the K-Dress from CuteCircuit, which contains LEDs [19]. The formation of companies focused on using electronic textiles as a craft skill to create bespoke garments has also resulted in a more general adoption of E-textile technology [20]. LED-yarns were used in the research described in this paper. This was a development on the history of inclusion of lighting within costumes for use in performance.

The aim of this research was to demonstrate the feasibility of attaching multiple LED E-yarns onto a stretch garment for use in performance. This showed the viability of fabricating clothing containing many electronic components held close to the skin of the wearer without impeding the wearer's ability to dance. Details are provided of a zig-zag E-yarn shape developed to accommodate the non-stretch E-yarn on stretch fabric. The design of the junctions between the E-yarns was found to be important, and development of these is discussed.

2. Materials and Methods

The choice of a carnival costume and competition, plus the aims described above, created a framework for the design process. The costume design focused around the use of E-yarns, with the method of attachment of numerous E-yarns onto the costume being key. The carnival "King" competition, and subsequent carnivals in which the costume was displayed, required large, eye-catching costumes in which a dancer can move fluidly, unrestricted by weight or bulk. For this large-scale carnival costume, much of the dancer's body would be covered, so that fabric next to the skin could not easily be seen in many areas of the costume. The LED-yarns were therefore included within the legs of the costume, as the legs would be moving and simultaneously visible to the audience during dancing.

The carnival costume involved two distinct parts: a unitard onto which the E-yarns were embroidered, and a frame worn by a dancer, to which many of the other parts of the costume were attached. The incorporation of LED-yarn into the legs of the costume was the focus of the research described in this paper. A decision was made to use a unitard onto which the LED-yarns would be attached as this provided a skin-tight, stretch platform for attachment of the LED-yarns. A black stretch unitard was chosen (Capezio® Men's Footless Tank Unitard in Black, Capezio®, Norwich, UK). Placing the LED-yarns on the legs of the costume ensured that they were visible and subjected to movement during dancing, testing the ability of the E-yarn to withstand movement while attached to a stretch fabric. A flame theme was chosen for the costume, providing the opportunity to use curved and illuminated flame-like patterns. The chosen pattern contained E-yarns in horizontal bands around the unitard legs, with interconnecting LED-yarns between parallel E-yarns that were connected to a battery in a pocket on the back of the unitard. The E-yarn was non-stretch due to the central strand of conductive copper within. A curving pattern was chosen for the E-yarns to allow the underlying fabric to stretch without breaking the E-yarn.

2.1. E-Yarn Attachment to Stretch Fabric

Initial test attachments of the E-yarn to stretch fabric were performed by hand, using a blind stitch (Figure 2a). The initial zig-zag design was slightly less controlled, but a zig-zag pattern with a more ordered structure was chosen for the final design as we believed that this would be easier to implement using an embroidery machine. We assumed that machine embroidery would form a strong attachment between E-yarns and fabric to withstand the stretch and movement of the fabric during performance. A sewing machine (Bernina 1000 Special, Steckborn, Switzerland) was used, with the E-yarn fed through a cording foot to create a wide zig-zag stitch to keep the yarn in place on the fabric surface. Figure 2b shows a sample of the stretch fabric with E-yarns machine embroidered into place in the chosen zig-zag pattern.

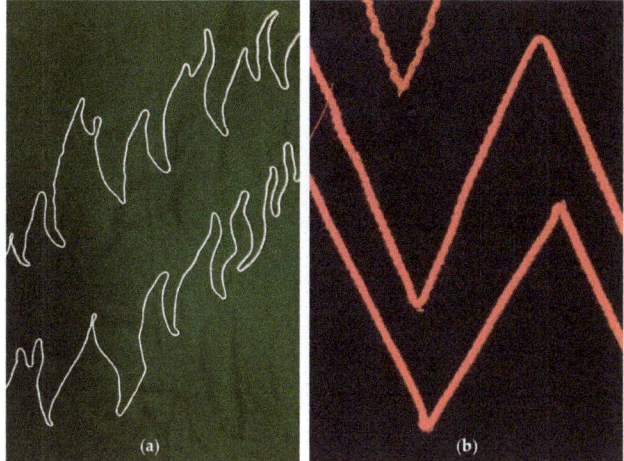

Figure 2. Zig-zag attachment pattern of electronic yarns (E-yarns) to stretch fabric: (**a**) Initial design, and (**b**) Final, more-ordered design.

2.2. Connecting E-Yarns

Connections between E-yarns, and between E-yarns and LED-yarns, were required to create an electrical circuit on the surface of the unitard. These were created by twisting together the copper wires protruding from the end of each E-yarn. These connections were then soldered together. Dymax 9001-E-V3.5 resin (Dymax Corporation, Torrington, CT, USA) was used to cover and reinforce the soldered joints. This was cured under UV light for 180 s using a Dymax Bluewave 50 Light Curing System (Dymax Corporation, Torrington, CT, USA). The connections were manually stitched into place on the fabric surface using a satin stitch to provide protection from external abrasion.

2.3. Tensile Testing of the E-Yarn Attachment to Stretch Fabric

Tensile testing was used to determine whether the curved pattern of the E-yarn attachment to the fabric surface was sufficient to prevent E-yarn breakage. Samples of stretch fabric (210 × 60 mm) were cut from a unitard identical to the one used to make the costume. Four of each sample were made, with the E-yarn embroidered onto the surface of the fabric in the following pattern:

- In a straight line.
- In a curved pattern.
- In a curved pattern, with a connection between the two E-yarns used to create the pattern.

The testing standard ASTM E8 [21] was the basis for tests on a zwickiLine tensile tester (Z2.5, Zwick/Roell, Ulm, Germany), using six testing cycles with a 30 s hold on the last cycle. The samples were taken to 50% strain, which was assumed to exceed the level of stretch to be experienced by the unitard to which E-yarns were attached. At this strain, the fabric was permanently damaged. Continuity testing was performed on each sample before and after tensile testing to find if the central conductive element of the E-yarn remained intact.

2.4. Circuitry

A circuit was required to power the LED-yarns. This is shown in the diagram in Figure 3. The periphery of the diagram shows long lengths of black E-yarn connected to a lithium polymer battery. Short lengths of LED-yarn were connected in parallel into this main circuit. The diagram shows connections to two lengths of LED-yarn, each containing 12 LEDs. Each LED (KPHHS-1005SURCK,

Kingbright, Taipei, Taiwan) required 2 V, so each section containing 12 LEDs required 24 V to be supplied by the battery. The LED-yarns were connected to short lengths of red E-yarn that added colorful detail to the costume. These E-yarns were then connected to the black E-yarns of the main circuit that ran down the outside of the costume legs.

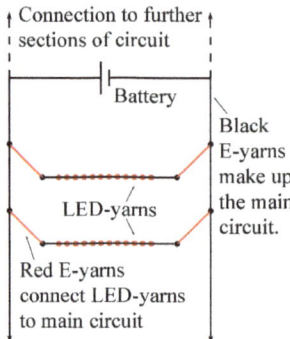

Figure 3. Diagram showing part of the circuitry attached to the costume: LED-yarns, each containing 12 LEDs attached in parallel between black E-yarns that form the main circuit on the costume. Red E-yarns connect the LED-yarns to this main circuit that is connected to a battery.

A section of the fabricated circuit is shown in Figure 4. This shows the zig-zag attachment pattern of the E-yarns and LED-yarns. A section of black LED-yarn containing 12 LEDs is highlighted in yellow. Each end of this highlighted LED-yarn was connected to a red E-yarn. The red E-yarns were placed above and below the LED-yarn to form a pattern on the costume. Three zig-zag lines of LEDs were included on the front of the costume legs. Ultimately, 144 LEDs were used, with two lengths of LED-yarn each containing 12 LEDs, placed at three levels on each leg.

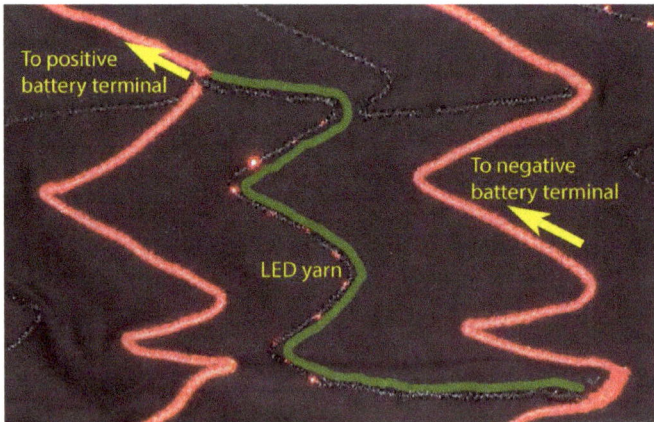

Figure 4. Circuitry on the costume: The LED-yarn is highlighted in green. Each LED-yarn contained 12 LEDs and was connected to two red E-yarns: One red E-yarn was connected to the positive terminal of a battery, and the other to the negative terminal.

The circuit was designed to incorporate the LED-yarns, with additional E-yarns connecting these to a battery placed on the center back of the unitard. The circuit was fabricated by placing the unitard on a mannequin to ensure that the costume legs would conform to the dancer's legs once the E-yarn was embroidered onto the fabric surface. Having placed the E-yarns in the desired positions, the circuit was created. Each positive and negative end of LED-yarn was connected to the E-yarns that were connected to the positive and negative battery terminals. The connections were soldered, coated with resin, and then embroidered into place as described in Section 2.2. The remaining lengths of E-yarn were machine stitched onto the unitard fabric using a zig-zag stitch. The machine sewing process proved to be too harsh, with almost half of the connections between the E-yarns breaking during the process. The broken connections were re-soldered and covered with resin. Sweat would possibly impact the performance of the E-yarns, causing conduction of electricity away from the lighting circuit. A resistor was included within the electrical circuit to ensure that no large power leakages could be experienced by the dancer. No current leakage was experienced by the dancer, but recommendations for further costume designs include use of insulated copper wire to minimise the possibility of current leakage.

2.5. Costume Frame

A carnival "King" costume must fill a large area to create an impact in competition and in parades [22], so the unitard with attached LED-yarns was worn underneath a backpack or frame, as shown in Figure 5. This was an aluminum structure to which rods were attached that held wings, flame shapes, extra LED lighting, and a "tail" of flexible rods. These radiated out from the center of the costume. The frame was made by an artisan specializing in large scale costume design [23]. Despite the costume size, the weight was reduced to a minimum through use of light, yet resilient, materials within and attached to the frame: composite rods, aluminum, lightweight fabrics, and foam sheet. The backpack was designed to make allowances for the placement of E-yarns and battery packs on the costume, without interfering with the LEDs' performance within the E-yarns, or with the electrical connections.

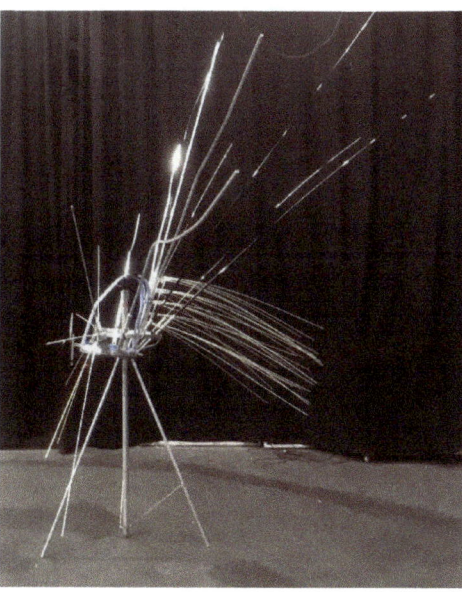

Figure 5. Completed frame to be worn over the unitard. The frame is shown supported on a stand.

3. Results

The carnival costume was successfully displayed in competition. The final costume design is shown in Figure 6a, with a detail of the illumination from the LED-yarns on the leggings in Figure 6b. A central design aim was to ensure that the performer was able to move and dance unimpeded and with confidence. This was achieved as the flexibility of the costume elements, including the E-yarns, allowed the whole costume to move smoothly as part of the choreography. Another design intention was that the LED-yarns should be visible. The level of illumination from the red LEDs on the unitard was minimal, with not all of the lines of the 12 LEDs illuminating. The E-yarns formed a very subtle detail amongst the overall impact of the costume, with higher levels of illumination from LED strips along the costume wings dominating.

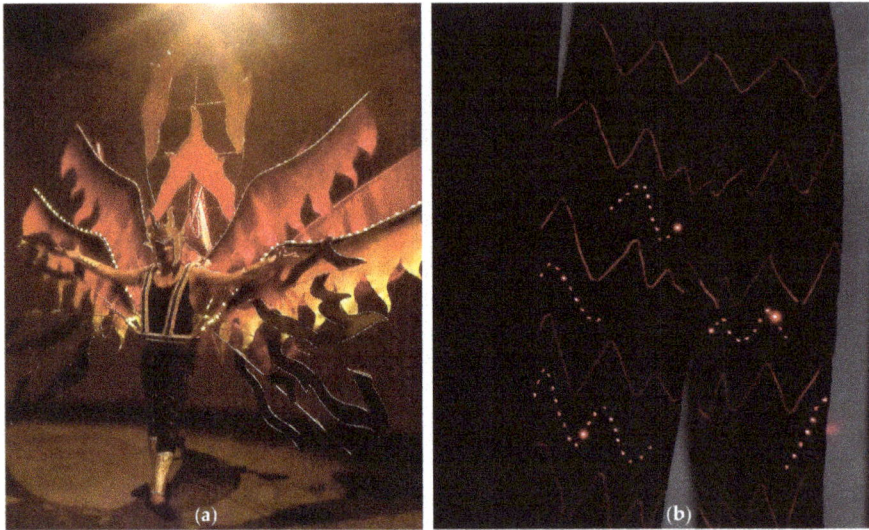

Figure 6. The completed costume. (**a**) The full costume incorporating the unitard, worn by Gil Santos. (**b**) A detail of the LED-yarns on the unitard legs, shown on a mannequin.

3.1. E-Yarn Connection Breakages

Tensile testing was performed to assess whether use of a curved E-yarn pattern of attachment to stretch fabric prevented breakage within the E-yarns. Examples of the tested samples are shown in Figure 7, and the results of tensile testing are provided in Figure 8. Each sample was tested over six cycles of increasing and decreasing strain. Breakages occurred during the first cycle of testing, so results from the first cycles of each test are displayed in Figure 8. The graph and continuity tests showed breakages in all of the samples containing a straight length of E-yarn. The samples containing a single length of curved E-yarn all remained intact, demonstrating that the curved placement of E-yarn on stretch fabric was effective at preventing E-yarn breakage. Ruptures occurred in all of the samples that incorporated a connection between two pieces of E-yarn. These breakages occurred at between 39% and 49% strain, as shown by the notches in the curves in Figure 8. The maximum stretchability of the fabric with attached, connected LED-yarns was therefore considered to be just below the lower measured strain value of 39%. This strain was considered to be well over the limit of strain to which the 10% Lycra® fabric would be subjected during performance. We therefore considered that the chosen methods of E-yarn placement, embroidery, and E-yarn connection were sufficient to ensure that the E-yarn circuitry remained intact.

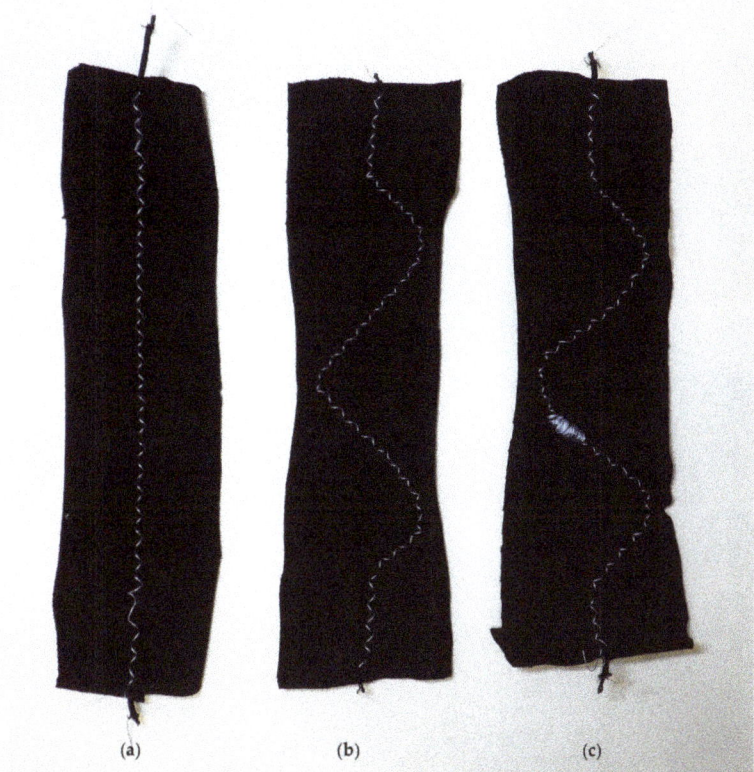

Figure 7. Samples of stretch fabric with attached E-yarns as prepared for tensile testing. From left: (**a**) Straight E-yarn; (**b**) Curved E-yarn; (**c**) Two curved E-yarns with a connection covered by a satin stitch.

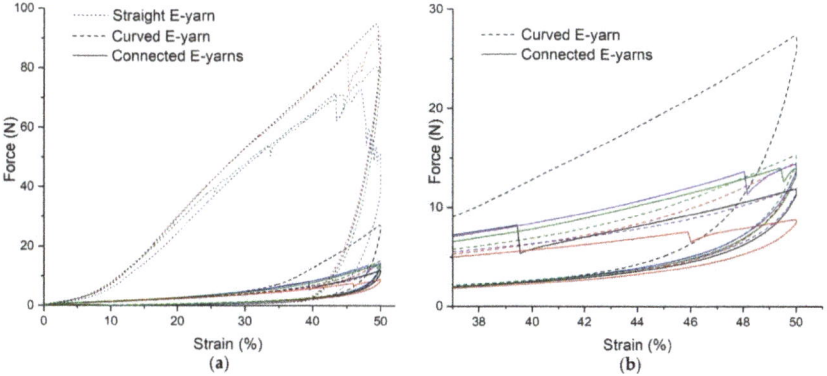

Figure 8. Results from the first cycles of cyclic tensile testing of the E-yarns attached to stretch fabric: (**a**) Notches near the top of the dotted curves show failure of straight E-yarns attached to stretch fabric; (**b**) The dashed lines in this detail from (**a**) show how E-yarns attached to the samples in a curved pattern remained intact. Notches in the solid lines show breakages in connections between curved E-yarns.

During performance, the E-yarns themselves remained intact, but many connections between E-yarns broke, causing the loss of illumination from some lines of LEDs that was apparent during the dance performance. We believed that these breakages happened when the unitard was pulled on by the performer, as the unitard underwent the greatest level of stretching during this process. The breakages occurred despite the assessment after tensile testing that connections between E-yarns would remain intact until beyond the level of strain at which the fabric would be permanently damaged. These breakages of connections also occurred despite the earlier work to reinforce the junctions between E-yarns after breakages occurred during attachment of E-yarns to the unitard.

The connections between the E-yarns were uncovered to check where the breakages had occurred and the weakest links were found to be the solder joints between wires. These areas were not completely covered with resin. This discovery was further confirmed after inspecting the broken links using a microscope (Keyence VHX-5000 Digital Microscope, Keyence (UK) Ltd., Milton Keynes, UK) as shown in Figure 9a. To correct this, the connections were re-soldered and then covered with a higher-viscosity resin (Dymax 9001-E-V3.7, Dymax Corporation, Torrington, CT, USA), ensuring that each solder joint was completely covered with resin.

Small areas of fabric were possibly subjected to high levels of strain at points where the E-yarn connections were embroidered into place. To prevent excess tension at the connecting points between E-yarns, we determined that longer lengths of E-yarn should be used to form zig-zags between connections. Extra lengths of black E-yarn were added in zig-zag patterns adjacent to connection points, as shown in Figure 9b.

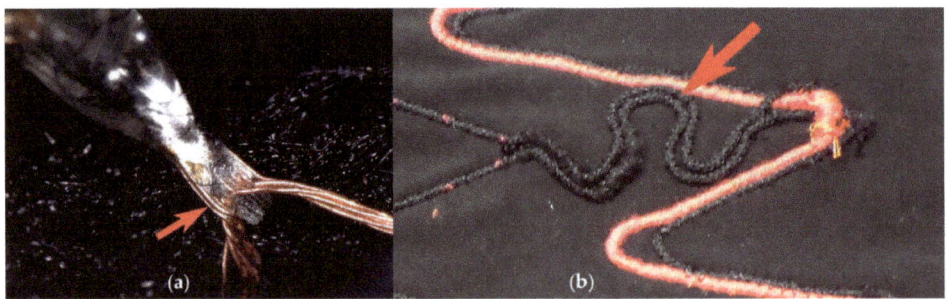

Figure 9. Breakage analysis of the E-yarns. (**a**) Connection between two copper wires magnified 100 times. The arrow indicates an area that is not covered by resin. (**b**) The arrow indicates a zig-zag connection between black LED-yarn and red E-yarn. This was added to reduce further breakages.

3.2. Washability

This prototype costume was not designed to be washed, but E-yarns in general are designed to be washable. Since creating the costume, initial wash tests have been completed. Five LED-yarns attached to clothing were machine washed and tumble dried. The LED-yarns still functioned after 7–25 washes, after which the test was stopped. Less aggressive methods of washing and drying, such as hand washing and line drying, are expected to lead to even greater longevity of LED-yarns. Future costumes containing E-yarns worn next to the skin can therefore be designed to be washable.

4. Discussion

This project demonstrated the flexibility of E-yarns, including the ability to attach E-yarns to skin-tight clothing without impeding the movement of a dance performer. The main hurdle for the use of numerous E-yarns on stretch fabric was found to be the creation of strong connections between the E-yarns. The method of forming the connections was improved during the fabrication of the costume, and afterward, but the process was time consuming. Each connection required soldering, application of resin, and couching in place while ensuring sufficient slack was left in the E-yarn on either side of the connection point. This was still not always sufficient to ensure that the connections remained unbroken during use. Ideally, the E-yarns themselves would be extendable, so that straight E-yarns could be attached to stretch fabric, without the need to zig-zag the E-yarn across the fabric surface. Further work is required to find a method of easily joining E-yarns easily to create robust junctions between flexible E-yarns, which can be relied upon to remain intact during use. This development will help ensure the feasibility and more widespread use of E-yarns.

LED Illumination

The illumination from the LEDs within the E-yarns was low, especially in contrast with the other LED lighting systems used on the upper parts of the costume, causing difficulties for audience members to see the LED-yarns that were key to the costume design. Ideally, the design would have been revised before the final assembly of the costume with white or lighter colored LEDs and a lighter or transparent textile on the outside of the E-yarns, resulting in more brightly-illuminated E-yarns. The preparation of the costume for a specific competition within the project timeframe did not allow for this modification. LEDs are best at visually demonstrating that E-yarns contain electronics, so a design containing brighter LEDs could be useful in demonstrating the existence of E-yarns.

The potential for much greater use of E-yarns lies in the range of electronic dyes and circuitry that can be incorporated within the E-yarns. LEDs provide a visible method of demonstrating this capability, so LED-yarns are important for their ability to catch the eye, drawing viewers in for further discussion about the technology.

5. Conclusions

The design of a carnival costume and its use in performance showed that electronic yarns (E-yarns) could be attached to stretch fabric for use in dance without impeding the movements of the dancer. This was the primary aim of our project. The LEDs within the LED-yarns functioned on the costume, although some electrical connections broke, resulting in a lack of illumination from some parts of the circuit. The low visibility of the red LEDs within the LED-yarns, plus the loss of some connections between the E-yarns, meant that a secondary design aim of demonstrating the existence of E-yarns to a wide audience was not completely fulfilled. Further work is required to create robust connections between E-yarns to ensure that the stretching of underlying fabric does not result in ruptured connections. Brighter LEDs within light-colored or clear LED-yarn would be more appropriate for use in a costume where higher visibility of the LEDs is required.

The design, fabrication, and testing in performance of this costume illustrate the potential for incorporating lighting into applications where flexibility is required, with the lighting as an integral part of a textile that does not impede the movement of the wearer. E-yarns can be designed to include many types of sensor, as well as LEDs, so the potential for integration of sensors into many types of stretchable clothing is illustrated.

Author Contributions: D.A.H., A.S. and T.H.-R. designed the study. D.A.H., V.S. and A.S. conducted the experiments. D.A.H. performed the data analysis. L.C. and V.S. designed the costume. A.M., V.S., L.C. and T.H.-R. fabricated the costume. A.M. provided specialist technical expertise. D.A.H. wrote the paper. All authors discussed the results and contributed to producing the final manuscript.

Funding: This research was funded by the Engineering and Physical Sciences Research Council under grant number EP/M015149/1 as a partnership project entitled "Use of electronic yarn in stretch fabric for the performing arts".

Acknowledgments: The authors would like to thank Steven Hoyte of Mas Rampage for construction of the costume frame; Carlos Oliveira, Ioannis Anastasopoulos, Mohamad Nour Nashed, and Achala Satharasinghe (of the Advanced Textiles Research Group, Nottingham Trent University), for technical assistance, especially in the fabrication and connection of E-yarns; Alice Cobbin and Evie Cobbin for assistance with the fabrication of the costume; and Tom Fisher and Kath Townsend of Nottingham Trent University for advice on the writing of this paper. The authors would also like to thank Gil Santos for dancing in the costume. Finally, the authors would like to express gratitude to Tatiana Woolley of ABC Dance School, Nottingham, for collaboration on the costume design, fabrication assistance, and organisation of dance performances.

Conflicts of Interest: The authors declare no conflict of interest. The founding sponsors had no role in the design of the study; in the collection, analyses, or interpretation of data; in the writing of the manuscript, and in the decision to publish the results.

References

1. Hayward, J. E-Textiles 2017–2027: Technologies, Markets, Players. Available online: https://www.idtechex.com/research/reports/e-textiles-2017-2027-technologies-markets-players-000522.asp (accessed on 8 February 2018).
2. Dias, T.K. Electronically Functional Yarns. World Patent WO2016/038342 A1, 17 March 2016.
3. Dias, T.; Hughes-Riley, T. Electronically Functional Yarns Transform Wearable Device Industry. *R D Mag.* **2017**, *59*, 19–21.
4. Hardy, D.A.; Anastasopoulos, I.; Nashed, M.N.; Oliveira, C.; Hughes-Riley, T.; Komolafe, A.; Tudor, J.; Torah, R.; Beeby, S.; Dias, T. An Automated Process for Inclusion of Package Dies and Circuitry within a Textile Yarn. In Proceedings of the Design, Test, Integration & Packaging of MEMS/MOEMS, Rome, Italy, 22–25 May 2018.
5. Rathnayake, A.S. Development of the Core Technology for the Creation of Electronically-Active, Smart Yarn. Ph.D. Thesis, Nottingham Trent University, Nottingham, UK, 2015.
6. Hughes-Riley, T.; Lugoda, P.; Dias, T.; Trabi, C.; Morris, R. A Study of Thermistor Performance within a Textile Structure. *Sensors* **2017**, *17*, 1804. [CrossRef] [PubMed]
7. EMCCAN "Carnival Clash" King and Queen Show; Theatre Performance at Nottingham Playhouse: Nottingham, UK, 2017.
8. Cork, C.; Dias, T.; Acti, T.; Ratnayaka, A.; Mbise, E.; Anastasopoulos, I.; Piper, A. The next generation of electronic textiles. In Proceedings of the 1st International Conference on Digital Technologies for the Textile Industries, Manchester, UK, 5–6 September 2013.
9. Guler, S.D.; Gannon, M.; Sicchio, K. *Crafting Wearables: Blending Technology with Fashion*; Apress: Berkeley, CA, USA, 2016; ISBN 1484218086.
10. Anonymous. The Electric Diadems of the New Ballet "La Farandole". *Sci. Am.* **1884**, *50*, 163.
11. Cook, J. How to Light Up Your Clothing. Available online: http://www.wired.co.uk/article/light-up-your-clothing (accessed on 1 November 2017).
12. Scottish Opera. The Magic Flute: Production Details. Available online: https://www.scottishopera.org.uk/for-hire/productions-for-hire/productions/the-magic-flute/ (accessed on 1 November 2017).
13. Anon Tron Dance—All about LED Dance Performance! Available online: http://www.trondance.com (accessed on 31 October 2017).
14. EL Wire Craft EL Wire and EL Tape. Available online: http://elwirecraft.co.uk/ (accessed on 1 November 2017).
15. Waldemeyer, M. Olympic Ceremonies—Moritz Waldemeyer. Available online: http://www.waldemeyer.com/olympic-ceremonies (accessed on 2 August 2017).
16. Moritz Waldemeyer Bono's Laser Jacket for the U2 360 Tour. Available online: http://www.waldemeyer.com/bonos-laser-jacket-u2-360-tour (accessed on 1 November 2017).
17. Pantouvaki, S. Embodied interactions: Towards an exploration of the expressive and narrative potential of performance costume through wearable technologies. *Scene* **2014**, *2*, 179–196. [CrossRef]
18. Anonymous. EFM: Apparel, Accessories, Advertisement. Available online: http://www.electroflashmedia.com/ (accessed on 1 November 2017).
19. CuteCircuit K-Dress—CUTECIRCUIT. Available online: http://shop.cutecircuit.com/products/k-dress-1?variant=267846634 (accessed on 12 December 2016).
20. Anonymous. About CuteCircuit. Available online: http://cutecircuit.com/about-cutecircuit/ (accessed on 1 November 2017).

21. *ASTM E8/E8M—16A Standard Test Methods for Tension Testing of Metallic Materials*; ASTM: West Conshohocken, PA, USA, 2016.
22. Achong, D. Mac Farlane Rules in Kings, Queens Prelims. Available online: http://www.guardian.co.tt/carnival/2012-02-11/mac-farlane-rules-kings-queens-prelims (accessed on 8 August 2017).
23. Mas Rampage Rampage Mas Band—Artists. Available online: http://rampagemasband.com/artists.html (accessed on 8 August 2017).

© 2018 by the authors. Licensee MDPI, Basel, Switzerland. This article is an open access article distributed under the terms and conditions of the Creative Commons Attribution (CC BY) license (http://creativecommons.org/licenses/by/4.0/).

Article

Developing Novel Temperature Sensing Garments for Health Monitoring Applications

Pasindu Lugoda [1,*], Theodore Hughes-Riley [1], Carlos Oliveira [1], Rob Morris [2] and Tilak Dias [1]

1. Advanced Textiles Research Group, School of Art & Design, Nottingham Trent University, Bonington Building, Dryden Street, Nottingham NG1 4GG, UK; theodore.hughesriley@ntu.ac.uk (T.H.-R.); jose.oliveira@ntu.ac.uk (C.O.); tilak.dias@ntu.ac.uk (T.D.)
2. School of Science and Technology, Nottingham Trent University, Clifton Lane, Nottingham NG11 8NS, UK; rob.morris@ntu.ac.uk
* Correspondence: Pasindu.lugoda2013@my.ntu.ac.uk; Tel.: +44-747-535-6784

Received: 30 April 2018; Accepted: 9 July 2018; Published: 10 July 2018

Abstract: Embedding temperature sensors within textiles provides an easy method for measuring skin temperature. Skin temperature measurements are an important parameter for a variety of health monitoring applications, where changes in temperature can indicate changes in health. This work uses a temperature sensing yarn, which was fully characterized in previous work, to create a series of temperature sensing garments: armbands, a glove, and a sock. The purpose of this work was to develop the design rules for creating temperature sensing garments and to understand the limitations of these devices. Detailed design considerations for all three devices are provided. Experiments were conducted to examine the effects of contact pressure on skin contact temperature measurements using textile-based temperature sensors. The temperature sensing sock was used for a short user trial where the foot skin temperature of five healthy volunteers was monitored under different conditions to identify the limitations of recording textile-based foot skin temperature measurements. The fit of the sock significantly affected the measurements. In some cases, wearing a shoe or walking also heavily influenced the temperature measurements. These variations show that textile-based foot skin temperature measurements may be problematic for applications where small temperature differences need to be measured.

Keywords: wearable electronics; wearables; smart textiles; electronic textiles; E-textile; digital medicine; temperature; thermistor; wound management; sensor network

1. Introduction

This work considers some of the practical aspects of on-body temperature measurements, with a focus on the creation of innovative temperature sensing garments. The use of textile temperature sensors allows for comfortable on-body temperature measurement, which is desirable for certain telemedicine and health monitoring applications. This work also presents a preliminary trial demonstrating a temperature sensing sock that was previously presented [1]. This trial creates new knowledge regarding skin temperature measurement of the foot using a textile-based temperature sensing garment.

Skin temperature is an important indicator of pathology and is one of the most commonly measured vital statistics in both infants [2] and adults [3]. Continuous temperature monitoring of the skin can provide clinicians with useful information when investigating a variety of conditions, such as non-freezing cold injuries [4,5], the early detection of foot ulcers [6,7], or Raynaud's disease [8].

Continuous remote temperature measurements can be provided through the use of wearable temperature monitoring devices. However, many wearable temperature monitoring devices are not easily concealed [9–11]. Patients prefer wearables that seamlessly integrate into their day-to-day lives while remaining hidden from view, as highlighted in recent studies [12].

Interest in electronic textiles has grown [13]. An electronic textile integrated with sensors is an ideal solution to provide truly discrete sensing by adding comfort and normalcy to the wearable device. A variety of temperature sensing textiles have been developed [14–17]; however, many are unable to provide localized (point) temperature measurements. The variation in skin temperature between certain points of skin is vital when monitoring conditions such as Raynaud's disease [8] and the early detection of foot ulcers [6,18,19].

Some textile solutions that allow for localized temperature measurements have been reported in the literature [20], including temperature sensing yarns fabricated using Electronic Yarn (E-yarn) technology [21,22].

This work focuses on using temperature sensing yarns, designed and characterized in earlier work [1,23–25], to create a series of innovative textile sensing garments. A key component of this work was to conceptualize and identify the feasibility of developing temperature sensing textiles using temperature sensing yarns. In order to integrate these temperature sensing yarns into garments, new knitting techniques had to be developed, creating important new knowledge. In addition to providing further details on the production of the temperature sensing sock [1], this work introduces a temperature sensing glove that could be used to create a temperature regulated glove for people with Raynaud's disease or those working in cold environments, and a temperature sensing armband for fever detection. The prototypes provide a fully textile-based solution to provide remote and continuous temperature measurements.

Full details are provided regarding the creation of the three prototype device types, including the knitting details and information about the associated interface hardware. The use of temperature sensing yarns for skin temperature measurements has been investigated. A temperature sensing yarn is initially used to measure the temperature at different points on a hand to better understand the nuances of recording skin contact temperature measurements with the yarns. A detailed experiment is then conducted to identify the effects of contact pressure on the temperature sensing yarn and how this affected the skin-based temperature measurements. The knowledge generated in this work will be useful for other textile-based temperature sensors. A preliminary trial is presented using the temperature sensing socks. This provided useful information about factors affecting this type of textile-based temperature measurement of the foot.

2. Materials and Methods

2.1. Temperature Sensing Yarn Fabrication

Temperature sensing yarns were produced using a method similar to that described previously [1,23–25]. To produce the yarns Murata 10 kΩ Negative Temperature Coefficient (NTC) thermistors (NCP15XH103F03RC; Murata, Kyoto, Japan) were soldered onto fine copper wires. The thermistor and interconnects were then encapsulated using an ultraviolet (UV) curable polymer resin with high thermal conductivity (Dymax 9-20801, Dymax Corporation, Torrington, CT, USA) to form a cylindrical micro-pod with a diameter of 0.87 mm and a length of 2.17 mm. The micro-pod was then covered with packing fibers and a warp knitted tube to create the temperature sensing yarns.

Previous work showed that the accuracy of the temperature-sensing yarns was ± 0.5 °C 63% of the time, or ± 1 °C 89% of the time [1]. This was regarded as acceptable since the accuracy of the thermistor specified by the manufacturer for the inspected range (22.25–62.15 °C) was ± 1.37 °C.

2.2. Prototype Temperature Sensing Armbands

Temperature sensing yarns were initially used to produce the armbands. The armband was chosen due to its tubular structure. The human body consists of structures with approximately circular cross-sections, hence it was important to identify if the temperature sensing yarn could be used to develop a tubular wearable device. Additionally, these armbands could be further developed into a fever detection device.

Three prototype armbands were produced, and for each of the prototypes, four temperature sensing yarns were used. The developmental process between each iteration of the armband is shown for completeness and to highlight the practical considerations for creating a monitoring device of this type. During development, the supporting hardware used to record measurements from the yarns was changed for each armband. Additionally, the connection method between the temperature sensing yarn and the hardware was altered in each iteration. For all prototypes, LabVIEW (National Instruments, Austin, TX, USA) was used to present the data captured. Figure 1 shows annotated photographs of each of the three armband prototypes.

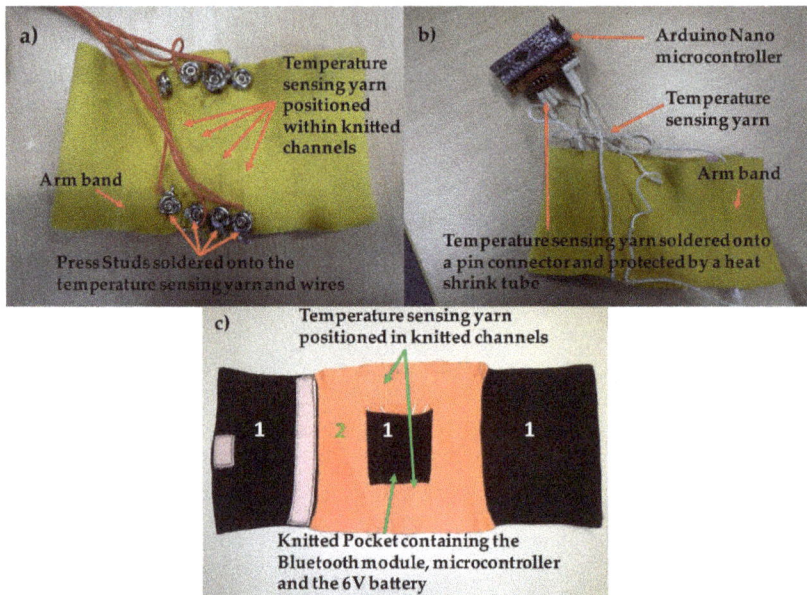

Figure 1. Annotated photographs of the three prototype armbands where (**a**) first armband prototype, (**b**) second armband prototype and (**c**) third armband prototype. The armbands were knitted with tubes to incorporate the temperature sensing yarns.

All armbands were knitted using a computerised flat-bed knitting machine, Stoll CMS 822HP E16 (Stoll, Reutlingen, Germany). A flat-bed knitting machine has precisely engineered needle beds composed of flat hardened steel plates. In industrial flat-bed knitting machines, a minimum of two such needle beds are arranged in an inverted V-form. In the needle beds, generally latch needles are placed inside needle tricks (open rectangular grooves precisely cut to accommodate needles) on the top surface of the needle bed. The above assembly guarantees the movement of needles individually and axially during the knitting process. A system of linear cams moves the needles between two dead centers in order to form stitches. The use of the needle-latch to open and close the needle hook area of a latch needle simplifies the stitch formation process.

Modern computerised flat-bed knitting machines are equipped with an electromagnetic system to facilitate the selection of individual needles during the knitting process. This combination of needle tricks, latch needles, and independent needle selection enables the creation of complex three-dimensional (3D) knitted structures on these machines. The use of two needle beds also provides the two needle systems with freedom of movement in two independent planes, thus forming the basis for forming stiches in 3D space.

As such, this technology was used to produce samples with temperature sensing yarns using two basic knitted structures integrated within the same knitted textile: the plain knitted structure and the interlock knitted structure.

2.2.1. First Armband Prototype

The first prototype (Figure 1a) was knitted using a double covered yarn made of 44F13 dtex lycra yarn that was covered with two yellow 78F46 dtex PA66, yarns (Wykes International Ltd., Leicester, UK). The double covered yarn was used due to its elasticity, ensuring that the final armband provided a snug fit. The base fabric was knitted in interlock and four 2 mm diameter tubes were knitted using the plain knitted structure. Four temperature sensing yarns were manually inserted into the plain knitted tube structures using a folded steel guide wire (diameter of 0.49 ± 0.01 mm). This yarn insertion technique was used to insert the temperature sensing yarns into all the prototype garments described in this paper. This manual technique was used due to the difficulty of manufacturing temperature sensing yarns in bulk quantities. Dias et al. have now semi-automated the manufacturing process for these E-yarns [26] so these yarns can be knitted using the Stoll flat-bed knitting machine. In the fabric produced for the first armband, the knitted channels were spaced 10 mm apart to accommodate temperature sensing yarns.

The temperature sensing yarns were connected to the interface hardware using a press-stud connection (press-stud diameter 7.53 mm) as these have been used in several electronic textile applications to form electrical connections [27,28] and create a strong mechanical and electrical connection. The temperature sensing yarns were soldered onto the female part of the metal press-studs as shown in Figure 1a. Thereafter, wires were soldered onto the male part of the press-stud, with the wires leading into a potential divider circuit, which contained 10 kΩ resistors that acted as the second resistors. The exact value of the second resistor was determined with a digital multimeter (Agilent 34410A, Agilent Technologies, Santa Clara, CA, USA) to a precision of 0.01%. The potential divider circuit was then connected to a data acquisition unit (NI DAQ USB 6008). Data were collected and interpreted using a bespoke LabVIEW script.

The main limitation when using this approach was the size of the NI DAQ USB 6008 unit (84.98 × 64.01 × 23.19 mm), which caused connection failures at the press-studs due to its weight. Another issue was caused by the large size of the press-studs, which prevented the temperature sensing yarns from being positioned in close proximity to each other in a fabric.

2.2.2. Second Armband Prototype

For the second armband (Figure 1b), the temperature sensing yarns were positioned 60 mm apart. For this prototype press-studs were not used. Instead, the temperature sensing yarns were soldered directly onto 20-mm-long male flat header connectors with a pitch of 2 mm. The solder joints were then encased within 2.4-mm-diameter heat shrinkable sleeves (Stock No. 397-4263, RS) to enhance the mechanical strength of the connection. An Arduino Nano v3.0 (Arduino, Turin, Italy) was used as the microcontroller instead of the NI DAQ USB 6008 unit due to its smaller size (43.18 × 18.54 mm). The Arduino Nano was then wired into a computer via a mini-B USB cable. The LabVIEW program was modified in order to read the data from the Arduino Nano.

2.2.3. Third Armband Prototype

The third armband (Figure 1c) was knitted using the technique described earlier; however, this design included an integrated pocket. As shown in Figure 1c, the third armband was knitted using two different types of yarns. The base fabric structure (shown by label 2 in Figure 1c) was knitted using a non-elastic 2/32 tex orange Merino wool yarn (Yeoman Yarns, Leicester, UK). The structure also contained four plain knitted tubes for the temperature sensing yarns. Two of the four plain knitted tubes were positioned 40 mm apart above the pocket and the remaining two were positioned 40 mm apart below the pocket. Non-elastic yarn was used to identify if this yarn would influence the snug fit of the armband and in turn, whether the fitting would affect the contact between the temperature sensing yarns and the skin. The sides of the prototype armband and the integrated pocket (labelled 1 in Figure 1c) were knitted using an interlock structure and a double covered yarn made of 44F13 dtex lycra yarn covered with two black 78F46 dtex PA66 yarns (Wykes International Ltd., Leicester, UK). This ensured that the sides of the armband could stretch to fit the wearer's arm.

This prototype armband could be connected to a PC wirelessly, with the interface hardware included in the knitted pocket of the armband. An Arduino Pro Mini (Arduino, Turin, Italy) was used as the microcontroller due to its small size (17 × 33 mm) and this was connected to a Bluetooth module from Sparkfun Bluetooth Mate Silver (SparkFun Electronics, Boulder, CO, USA) to provide wireless connectivity. This Bluetooth module was chosen due to its low power consumption; however, this also limited its transmission range (the Sparkfun Bluetooth Mate Silver used a RN-42 class 2 Bluetooth module).

The main problem experienced when using the third prototype armband to obtain temperature measurements was the random drop in the Bluetooth signal. It was also observed that the Merino wool yarn failed to provide proper contact between the temperature sensing yarns and the skin; the inadequate stretch properties of Merino wool meant that the armband fitted loosely on the arm.

The experience of creating the temperature sensing armbands and their performance informed the design of two further prototypes: the temperature sensing glove and temperature sensing sock. Therefore, details of the armbands are included in this work for completeness only. Additional experimental work was not conducted using the armband designs presented here.

2.3. Prototype Temeprature Sensing Glove

The prototype temperature sensing glove was developed using the Stoll computerised flat-bed knitting machine described earlier. The prototype was knitted as a seamless glove with integrated tubes for inserting the temperature sensing yarns using double covered yarn composed of 44F13 dtex lycra yarn covered with two black PA66, 78F46 dtex PA66 yarns.

The glove was created by forming successive courses (rows of loops) parallel to the main glove axis along the line of the middle finger. The first row of stitches was formed along the line of the smallest finger, with the process continuing to successively form the four fingers and the thumb with the core section (body of the glove). Each finger was knitted from its distal to its proximal end, with the proximal ends then linked to form a core section. The fingers were knitted using a "C-knitting" process. The thumb was knitted with the core section merging with the proximal end of the thumb, which was continued to complete the glove. By using the C-knitting process for the thumb in this method, the overall shape of the glove could be adapted to the natural shape of the human hand. The fingers were finished by binding the last rows of knitting on the two needle beds together in order to create a seamless glove with a better fit.

Five tubular channels were integrated within the knitted glove structure to accommodate the temperature sensing yarns and to ensure that they remained hidden. This guaranteed that the aesthetics of the glove were not affected by the temperature sensing yarns. After knitting, the five temperature sensing yarns were incorporated into the tubular channels with the sensing elements (thermistors) of the temperature sensing yarns positioned at the tips of the five fingers.

The interface hardware design used for Armband Prototype 3 (Figure 1c) was also used for the glove. However, instead of the Sparkfun Bluetooth Mate Silver, a Sparkfun Bluetooth Mate gold module was used, which increased the range of transmission at the cost of consuming more power. The interface hardware was powered using two coin CR2025 3V Lithium Coin Cell (Maplin Electronics, Rotherham, UK). The batteries, the Arduino Pro Mini, and Sparkfun Bluetooth Mate gold module were stacked on top of each other and positioned at the back of the wrist.

A bespoke LabVIEW script was used to provide the user interface. A picture of a hand was included on the front panel of the program with colored boxes placed on each finger at the position of the five thermistors in the glove. This allowed the user to set temperature limits so that once the temperature went above the set limit, the box turned red, or when it went below, the box turned blue; otherwise, the box remained green. The user interface for the glove and the prototype glove are shown in Figure 2.

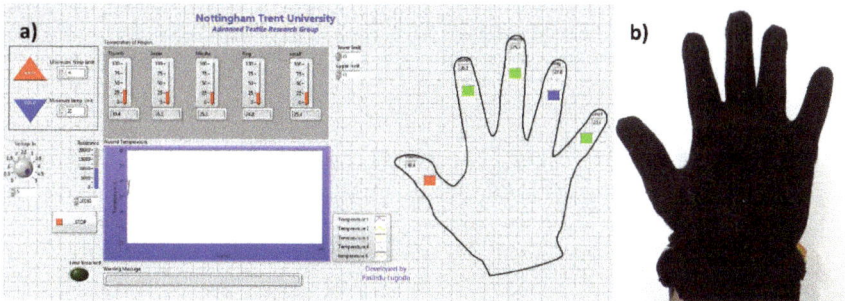

Figure 2. Prototype temperature sensing glove. (**a**) The LabVIEW software user interface for the glove. (**b**) A photograph of the prototype glove.

2.4. Prototype Temeprature Sensing Sock

The experience gained by producing the temperature sensing armbands and temperature sensing glove was used to improve the design of a temperature sensing sock. Five temperature sensing yarns were used to produce a sock that could detect temperature at five different points on a foot. The sock was previously and briefly described elsewhere in the literature [1]. The sock was produced using a computerized flat-bed seamless knitting machine (Model SWG 091N3, E15, Shima Seiki, Sakata Wakayama, Japan). The knitted structure had five tubular channels to incorporate the temperature sensing yarns, similar to the technique described in the previous section. A 100% combed black 3/42 tex cotton yarn (Yeoman yarns, Leicester, UK) was used to manufacture the sock. Cotton was chosen as it is one of the most commonly used materials in the manufacture of socks and would therefore add normalcy to the final prototype.

As with all knitted materials, the knitted structure relaxes and shrinks in size after manufacturing and may stretch when worn. Therefore, to ensure that the sensing elements of the temperature sensing yarn were positioned correctly when the sock was worn, a simulated foot was created using plaster (Gypsum). The five positions chosen were the big toe, heel, and three points on the metatarsal head to provide a good indication of the temperature across it. Metal studs were integrated onto the plaster at the five chosen locations to help position the sensing elements of the temperature sensing yarns at the desired locations on the foot. After knitting the sock, the sock was placed onto the simulated foot and the sensing elements of the temperature sensing yarns were positioned to the precise location indicated by the metal studs.

The temperature sensing yarns were connected to a potential divider circuit as discussed earlier. The potential divider circuit was then connected to a USB 6008 DAQ unit that was interfaced to a computer using a USB 2.0 cable (type A to B, National Instruments). The computer provided the power required for the USB 6008 DAQ. The LabVIEW software developed for the glove was modified to be used with the sock. On the front panel, instead of having an image of a hand, an image of a foot was used as previously shown [1]. At the location of each of the sensors in the knitted sock, intensity graphs were positioned as indicators on the foot image. Intensity graphs were used instead of color boxes as this provided a gradual change in color with the change in temperature, providing more information to the end user.

2.5. Measuring Skin Temperature of the Hand Using a Temperature Sensing Yarn

It was important to understand the behavior of the temperature sensing yarn when taking skin contact temperature measurements and to validate that the temperature sensing yarn still operated correctly. Therefore, a preliminary experiment was performed using healthy volunteers from the research team to observe the effects of using the temperature sensing yarn for temperature measurements at different points on the hand.

The temperature sensing yarn was placed at different positions on the hand (side, palm, back) as shown in Figure 3. Two 100 g weights were attached to the temperature sensing yarn to create uniform tension. The resulting tension on the yarn (T_{hold}) can be calculated using the Capstan equation [29]:

$$T_{load} = T_{hold} e^{\mu \beta} \quad (1)$$

where T_{load} is the tension applied by the weights, μ is the coefficient of friction between the temperature sensing yarn and the skin, and β is the total angle swept by all turns of the temperature sensing yarn.

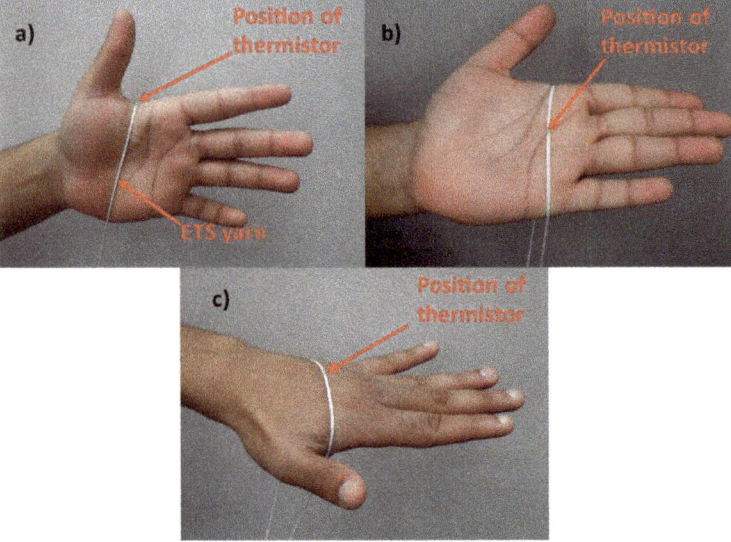

Figure 3. Positioning the temperature sensing yarn: (**a**) side, (**b**) palm, and (**c**) back of the hand.

Three positions on the hand were investigated, with each of these positions providing different contact surfaces for the temperature sensing yarn. In order to validate the measurements, two k-type thermocouples (Pico Technology, St Neots, UK) were positioned on the skin at either side of the temperature sensing yarn. Additionally, a Raytek Raynger MX Infrared Thermometer (Raytek® Fluke Process Instruments, Santa Cruz, CA, USA) was also used to obtain non-contact temperature measurements. This system had an accuracy of ±1 °C.

Experiments were conducted on three separate healthy volunteers from the research team. For each volunteer, measurements were recorded using three different temperature sensing yarns, with each yarn placed at the three positions on the hand, as shown in Figure 3. The temperature sensing yarn was maintained in each position for two minutes (44 measurements were obtained each minute) and the average temperature measurements during the last 30 s were used as the final temperature. This was done to ensure that a steady state was reached before the measurements were taken. Previous work has shown that a temperature sensing yarn has a step-response time of 0.17 ± 0.07 s while heating [1].

To record the temperature measurements, the temperature sensing yarns were connected to a potential divider circuit as described earlier, which was then connected to a data acquisition system (NI DAQ USB 6008, National Instruments). The resistance values recorded using the DAQ were converted to temperature values using the conversion equation provided by the thermistor manufacturer. The thermocouples were connected to a thermocouple data logger (PICO-TC08, Pico Technology, St Neots, UK). LabVIEW was used to capture the temperature from the temperature sensing yarns and the thermocouples.

2.5.1. Effects of Increasing Contact Pressure on Measurements Recorded Using the Temperature Sensing Yarns

The temperature sensing yarn and the hand are both 3D structures that deform under pressure. Therefore, identifying the effects of contact pressure on measurements recorded using the temperature sensing yarns was important. In order to achieve this, the same experimental procedure mentioned above was used. However, the temperature sensing yarn was only positioned at the side of the hand (as shown in Figure 3a) and the two weights attached to the temperature sensing yarn were varied (10, 20, 40, 100, 200, and 400 g), changing the contact pressure between the yarn and the hand. Readings from the temperature sensing yarns, the thermocouples measuring skin temperature, and a thermocouple measuring room temperature were recorded, as previously discussed.

The data are presented as a measurement error. The measurement error is the difference between the surface temperature and the temperature indicated by the sensor. For these experiments, the true surface temperature was assumed to be the temperature captured by the thermocouples positioned on either side of the hand. The relationship between the true surface temperature and the indicated temperature can be defined using Equation (2) [30]:

$$Z = \frac{(T_s - T_i)}{(T_s - T_a)} \qquad (2)$$

where Z is the measurement error, T_s is the true surface temperature, T_i is the indicated temperature, and T_a is the room temperature.

2.5.2. Measuring the Temperature of a Rigid Surface Using the Temperature Sensing Yarn

It was crucial to identify the effects of contact pressure on the temperature sensing yarn measurements when it was used to measure a rigid surface as, unlike the hand, the rigid surface does not deform with increasing contact pressure. The following experiments were conducted using the temperature sensing yarns and a Weller WS 81 (Mfr. Part No. T0053250699N, Weller®, Besigheim, Germany) soldering station. The temperature sensing yarn was placed over the shaft containing the solder tip (henceforth referred to as the shaft; diameter 6.77 mm) of the soldering iron with the soldering tip removed. Two weights were hung from either end of the temperature sensing yarn. The soldering iron was set to 150 °C with the shaft temperature at the point of measurement being recorded at 65.32 ± 3.80 °C using a k-type thermocouple (Pico Technology). Six weights (10, 20, 40, 100, 200, and 400 g) were used for these experiments. A k-type thermocouple (Pico Technology) was held onto the shaft using 3M™ Temflex™ 1300 vinyl electrical tape (3M, Maplewood, MN, USA), which was positioned about 1 mm away from the temperature sensing yarn. The room temperature was obtained using another k-type thermocouple (Pico Technology). Temperatures were captured and recorded using the method discussed in Section 2.5.

2.6. Preliminary User Trials Conducted on the Prototype Temperature Sensing Sock

In order to test the prototype temperature sensing sock, it was decided to evaluate how the temperature measurement from the sock varied under different operational conditions. When the sock was not worn, the sock was worn, when a shoe is put on, and when stepping while wearing a shoe were all investigated. These tests were conducted using five healthy volunteers from within the research team. Initially the socks were put on to the simulated foot for five minutes. This ensured that all five temperature sensing yarns were placed in the same environmental conditions and enabled a baseline reading to be recorded. Thereafter, the socks were worn by the volunteers and the temperature measurements were obtained over a period of five minutes. Next, a shoe was worn by the user and the temperature was obtained for another five minutes. Finally, the two feet were moved up and down on a step to simulate the effects of walking, during which time temperature measurements were recorded. Images of the shoes worn by the volunteers are shown in Figure 4.

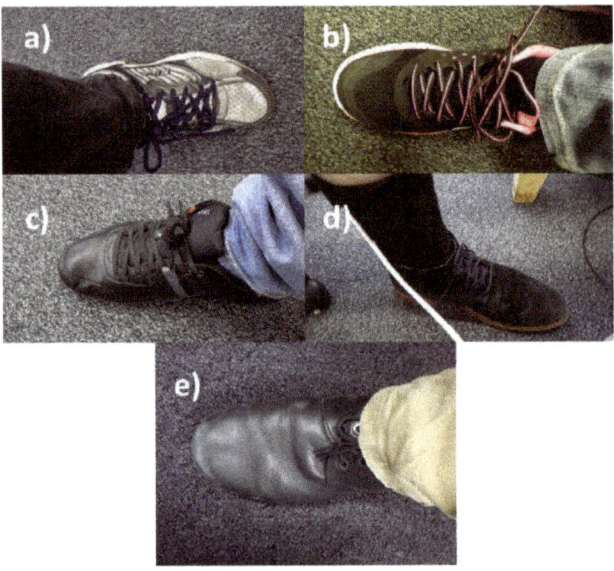

Figure 4. The shoes worn by the five volunteers for the trial. Different types of common footwear were used. The shoes worn by (**a**) Volunteer 1, (**b**) Volunteer 2, (**c**) Volunteer 3, (**d**) Volunteer 4 and (**e**) Volunteer 5.

An additional experiment was conducted on Volunteer 5, where a second cotton sock that was made using the same material as the first sock was worn on top of the sock containing temperature sensing yarns. This was done to better understand if an additional sock would enhance the contact between the temperature sensing sock and the skin, or if the additional layer of insulation would dramatically affect the results. Initially, measurements for when the temperature sensing sock was worn were captured. Then temperature measurements after the additional sock was worn on top of it were recorded.

3. Results and Discussion

3.1. Hand Temperature Sensing Yarn Measurements

Experiments were initially performed to validate the use of the temperature sensing yarns for skin contact measurements. Measurements were recorded with three volunteers using three different temperature sensing yarns at three different positions on the hand: side, palm, and back. The measurements from the three temperature sensing yarns were averaged and the standard deviation was calculated at each position for each of the volunteers. The average temperature measurements captured by the thermocouples and the infrared thermometer were also recorded. The difference between the thermocouple temperature measurements and that of the temperature sensing yarns were calculated and are shown in Figure 5. The recorded room temperatures (the room temperature was recorded using another k-type thermocouple) are also presented.

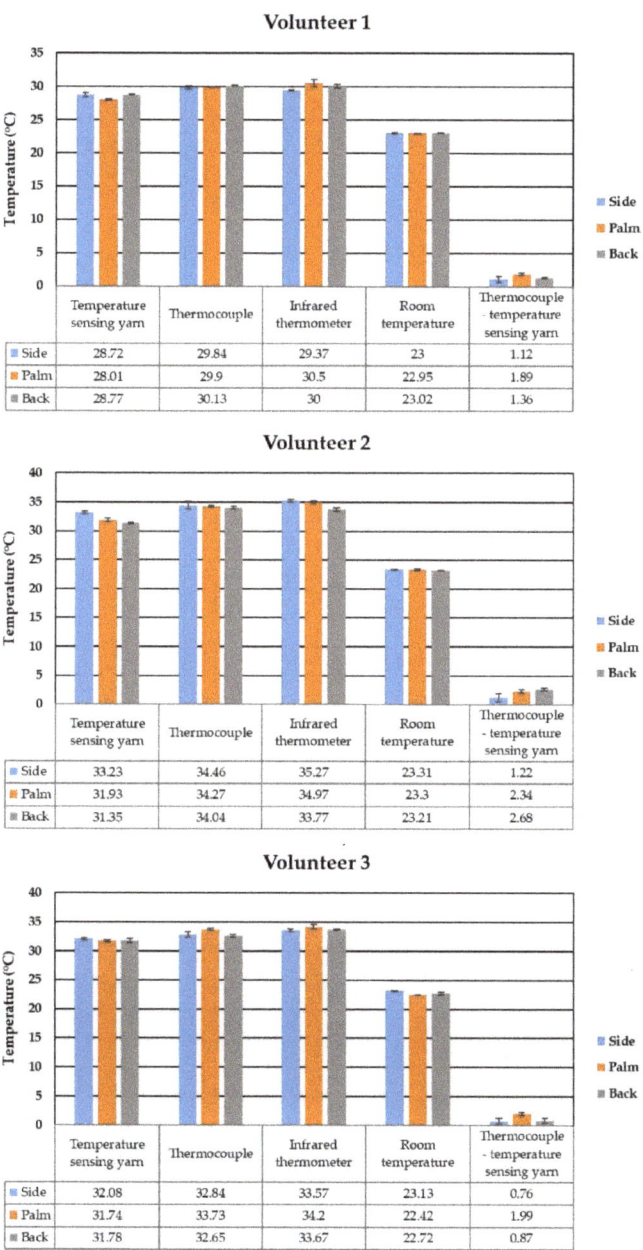

Figure 5. The measurements from the temperature sensing yarn, thermocouple, and infrared thermometer at different positions on the hand for the three different volunteers. The room temperature and difference between the thermocouple and yarn measurements are also presented. (**Top**) On the palm. (**Middle**) On the back of the hand. (**Bottom**) On the side of the hand.

From the results shown in Figure 5, the average temperature difference between the temperature sensing yarn measurements and the thermocouple measurements for each position from all of the three volunteers were observed to be 2.1, 1.6, and 1.0 °C for the palm, back, and side of the hand, respectively. These results illustrated that the difference between the temperature sensing yarn measurements and the thermocouple measurements were the largest when the temperature sensing yarns were positioned on the palm of the hand. This difference was minimized when the temperature sensing yarns were positioned on the side of the hand. This was potentially due to the lower radius of curvature at the side of the hand when compared with the other positions, which provided a higher surface contact pressure between the temperature sensing yarn and skin. According to the Laplace pressure equation in Equation (3) [31], when taking temperature measurements:

$$p = 2\sigma/r \tag{3}$$

where p is the pressure, σ is the surface tension, and r is the radius of curvature.

The temperature sensing yarn had a minimal contact pressure with the skin when it was placed on the palm of the hand, since the palm had the largest radius of curvature, which produced the largest difference in temperature readings as expected.

3.1.1. Effects on Measurements by the Temperature Sensing Yarn with Changing Contact Pressure

Temperature measurements of the hand were recorded for different contact pressure between the yarn and the hand. This was achieved by changing the weights attached to the yarn. The yarn was positioned on the side of the hand (as shown in Figure 3a) for all experiments. The measurement error was calculated between the surface contact temperature measurement using thermocouples and the temperature sensing yarn, as shown in Figure 6.

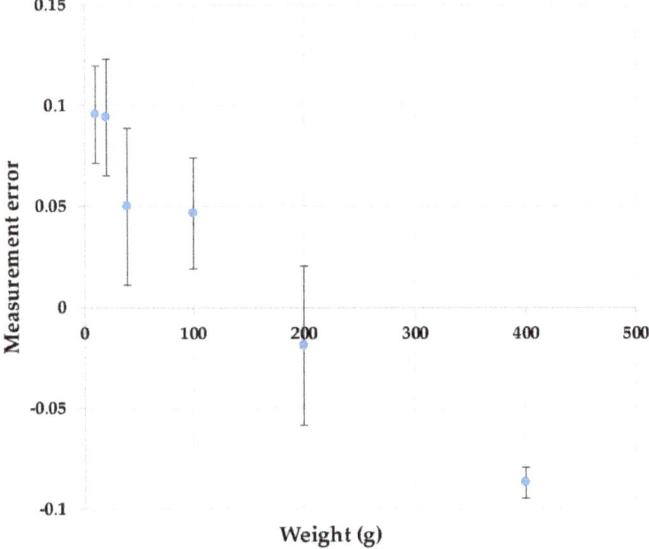

Figure 6. The change in the measurement error between the surface skin temperature and temperature sensing yarn when different weights were attached to the temperature sensing yarn, varying the contact pressure between the yarn and the hand.

As illustrated in Figure 6, the measurement error decreased with increasing weight and therefore the contact pressure. Relatively high experimental errors were observed mainly due to the difficulty of repositioning the yarns on the hand between experiments. The decrease in measurement error with an increase in contact pressure was due to the contact pressure deforming the 3D shape of the hand as well as the temperature sensing yarn. We also observed in the experiment where the 200 g and 400 g weights were used that the measurement error became negative (−0.047 and −0.090, respectively). Notably, the true surface temperature was obtained using the thermocouples that were just positioned on the surface of the skin. When the weight on the temperature sensing yarn was increased, the yarn began to sink into the skin. This increased the contact surface area between the skin and the yarn, which would increase heat transfer from the hand and restrict heat flow out of the yarn.

As the effect of contact pressure instigated the deformation of both the yarn and the hand, additional experiments were conducted using a rigid surface to better understand the effects of contact pressure on only the yarn.

3.1.2. Effects of Contact Pressure on the Measurements when Measuring a Rigid Surface

A measurement error was calculated between the surface contact temperature measurement of a rigid object (soldering iron shaft) and the temperature sensing yarn. Measurements were recorded for different weights attached to the temperature sensing yarn, varying the contact pressure between the yarn and the shaft, as shown in Figure 7.

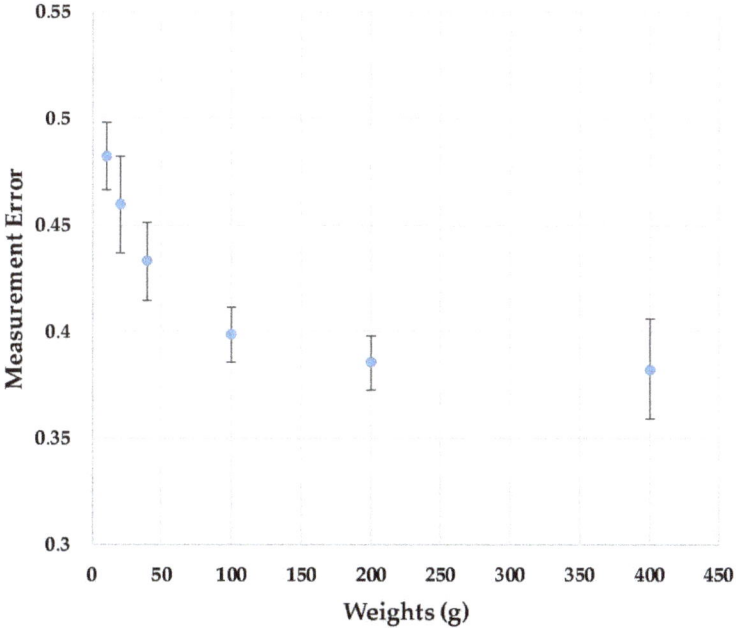

Figure 7. The change in the measurement error between the rigid surface (soldering iron shaft) temperature and the temperature sensing yarn when different weights were attached to the temperature sensing yarn, varying the contact pressure between the yarn and the shaft.

From Figure 7, increasing the weight (and therefore contact pressure) decreased the measurement error between the soldering iron shaft's surface temperature measured using a thermocouple and the temperature sensing yarn reading. However, unlike the previous experiment (Figure 6), the values did not become negative. This was a result of the hard metal surface of the shaft not deforming as a consequence of the increasing contact pressure. The decrease in the measurement error with increasing weights was likely due to the compression of the packing fibers and the fibers of the warp knitted tube with the increase in contact pressure. This would expel the air in the gaps between the knitted loops and increase the heat transfer from the soldering iron shaft and the temperature sensing yarn. The sensing element in the temperature sensing yarn would also be positioned closer to the shaft surface, further increasing heat transfer.

3.1.3. Discussion

Investigating temperature measurements using a temperature sensing yarn on the hand showed that the contact pressure had an effect on the measurements. When the temperature sensing yarn was used to capture temperature, contact pressure and the radius of curvature would be important factors in some cases. It is highly unlikely that these temperature sensing yarns would be used on their own to measure temperature; they would be inserted into a textile garment and then used as a tool to measure temperature. In certain situations, the effects of contact pressure would be less relevant. If the temperature sensing yarns were used as a tool to compare the skin temperature at discrete points on two different feet, for example, the radius of curvature would most likely remain similar. Hence, this effect could be ignored. In other cases, the garment and the fabric could be engineered to control the radius of curvature or to generate a known pressure, further minimizing this effect. Once these fabrics are engineered, a larger user trial should be conducted on these fabrics to understand the effects of contact pressure.

3.2. Measurements with the Prototype Temperature Sensing Sock

Experiments were conducted with the temperature sensing sock on five volunteers in three different scenarios: worn, worn under a shoe, and worn under a shoe while walking. The sock was also tested on a simulated foot before the experiments. An additional condition was also evaluated on Volunteer 5, where a second cotton sock was worn on top of the temperature sensing sock. These experiments were completed to understand the limitations of obtaining foot skin temperature measurements with a textile-based temperature sensors. Results from each of the test conditions are detailed below.

3.2.1. Sock Worn on the Simulated Foot

The results for when the sock was not worn by the volunteers but instead worn on a simulated foot are presented in Figure 8. Capturing and presenting the ambient temperature was essential, since ambient temperature has an impact on skin contact temperature measurements as shown in the literature [32]. These experiments were not conducted in a climate controlled room and therefore daily variations had to be understood when comparing volunteer data.

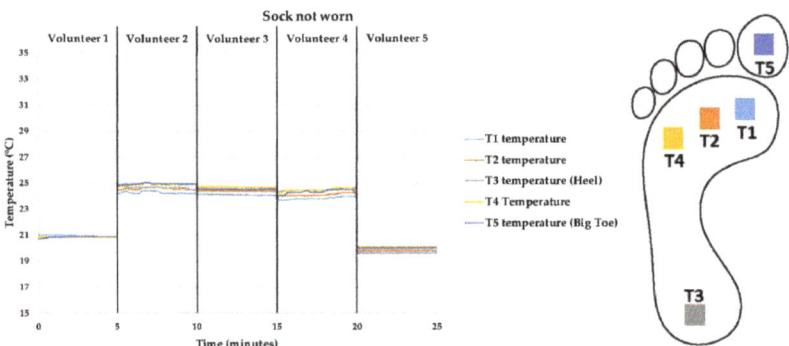

Figure 8. Temperature measurements from the temperature sensing sock when the sock was not worn by a volunteer. Temperatures were recorded prior to testing on all the volunteers.

Figure 8 shows the minor variations of >1 °C between the temperature measurements of the five temperature sensing yarns. This variation falls within the ±1.37 °C accuracy of the thermistor, which has been specified by the thermistor manufacturer. Notably, the user trials were not performed on the same day; therefore, as seen in Figure 8, the ambient temperatures recorded prior to testing on each volunteer were different. Regardless, in the cases of the second, third, and fourth volunteers, the ambient temperatures captured were within a ±1 °C and this provided similar test conditions for these three volunteers.

3.2.2. Sock When Worn by Volunteers

Trials were subsequently performed with the temperature sensing sock worn by volunteers. The temperatures captured by the temperature sensing yarns in the sock when the sock was worn by each of the volunteers are presented in Figure 9.

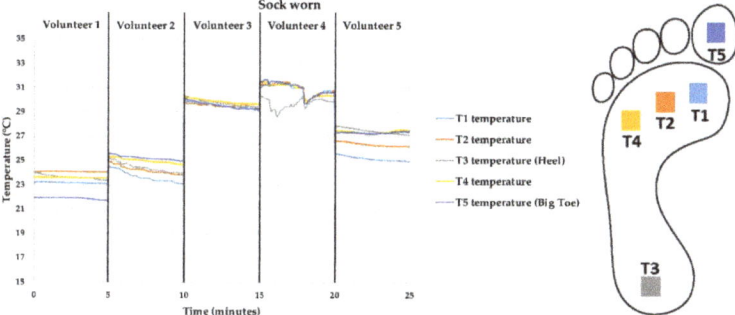

Figure 9. Temperature measurements from the temperature sensing sock when the sock was worn by each of the volunteers.

From Figure 9, when the sock was worn, the temperature increased for all the volunteers. For all volunteers except for Volunteer 1, the foot temperature measurements gradually decreased with time, which may have been caused due to cooling bought about by the room temperature.

For the first two volunteers, the temperature measurements were below 28 °C, which was outside of the standard deviation of the mean measured foot temperatures presented in the literature (30.6 ± 2.6 °C) [33]. This may have been due to poor contact between the sock and the foot. The socks were made for a UK size 9 foot but the feet size of the first two volunteers were size 7 and 5, respectively. The temperature recorded from the next two volunteers were within the mean measured foot temperatures presented in the literature. This was most likely due to the sock fitting well on the wearer's feet. The foot size of the third and fourth volunteers were 8.5 and 9, respectively. For the fourth volunteer, the heel temperature (T3) was notably lower during the first three minutes compared with the other temperature sensing yarns. This may have been caused by improper contact between the temperature sensing yarn and the heel. We also observed a significant fall in the temperature readings after the first three minutes for the fourth volunteer. This might have been due to the volunteer changing the position of their foot during the experiment.

The fifth volunteer had size 11 feet, implying that the temperature sensing sock would have had a close fit. The temperatures obtained from the fifth volunteer were not within the expected range (30.6 ± 2.6 °C). The fit of the sock might have affected the positioning of the thermistors in the temperature sensing yarns (especially in the cases of T1 and T2) and this could have led to improper contact between the foot and the temperature sensing yarns. The lower ambient temperature observed in the case of the fifth volunteer (Figure 8) would have also lowered the temperatures captured by the temperature sensing yarns in the sock.

Therefore, in order to obtain highly accurate or relative measurements, the socks have to be well fitted to the wearer's feet.

3.2.3. Sock Worn with a Shoe

Temperature measurements recorded by the temperature sensing yarns from all the volunteers when a shoe was worn on top of the sock are provided in Figure 10.

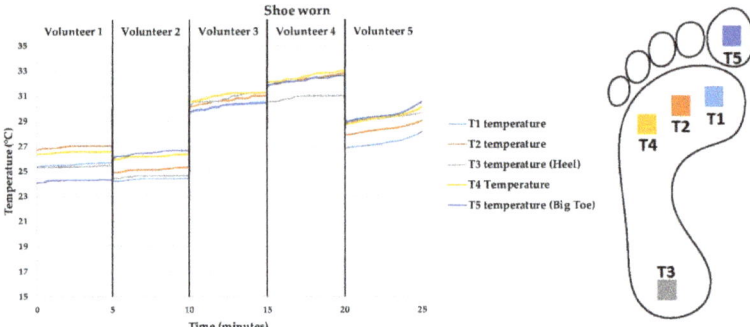

Figure 10. Temperature measurement from the temperature sensing sock when the sock was worn under a shoe for all of the volunteers.

The temperatures recorded increased for all the volunteers when a shoe was worn with the sock. The shoes provided insulation against the lower ambient temperature and reduced the heat flow from the foot to the atmosphere. The most significant rise in the temperature measurements was observed in the fourth volunteer who wore a boot, where the temperature rose by 1.5 °C in four of the five sensors (T1, T2, T4, and T5). This may have been caused by the extra insulation provided by the boot.

The data clearly showed that the shoe anatomy also impacted the temperatures captured by the temperature sensing yarns in the sock. Therefore, it can be concluded that wearing a shoe over the sock alters the temperature measurements captured by the sock and that the anatomy of the shoe may also have an impact on the measurements.

3.2.4. Shoe and Sock Worn While Walking

The results from walking while wearing the temperature sensing sock and a shoe for all the volunteers is presented in Figure 11.

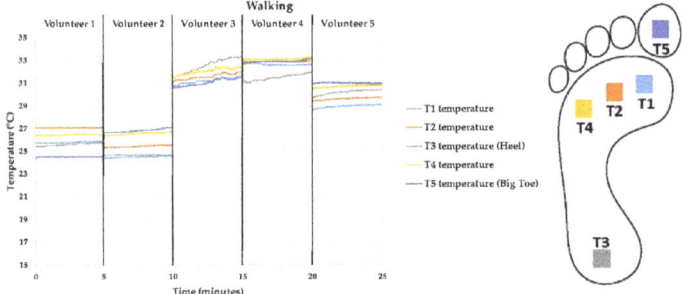

Figure 11. Temperature measurement from the temperature sensing sock when the sock worn under a shoe while walking for all the volunteers.

As illustrated in Figure 11, walking caused dissimilar changes in the temperatures recorded for each of the volunteers. For Volunteer 1, the temperature measurements were the same when the feet were moving compared to when they were stationary. For Volunteer 2, a minimal increase occurred in the temperature recorded by four temperature sensing yarns (T1, T2, T4, and T5) when the user started walking. The temperature of the foot rose the most for Volunteer 3. This may have been due to the fact that the volunteer moved their feet faster than the rest of the volunteers. This resulted in a difference up to 3 °C compared to when only the sock was worn (i.e., no movement and no shoe worn). In the case of Volunteer 4, motion resulted in a small increase in temperature. For Volunteer 5, the temperature recorded by four of the temperature sensing yarns (except for the sensor on the big toe, T5) increased slightly. Therefore it was concluded that walking with these temperature sensing socks caused changes in the temperatures recorded. The changes in temperature depend on the individual wearing the sock as well as the walking speed; however, further experiments are needed to fully quantify this effect.

3.2.5. Wearing an Additional Sock over the Temperature Sensing Sock

The results for when Volunteer 5 wore the temperature sensing sock and when an additional sock was worn on top of the temperature sensing sock were recorded and are shown in Figure 12.

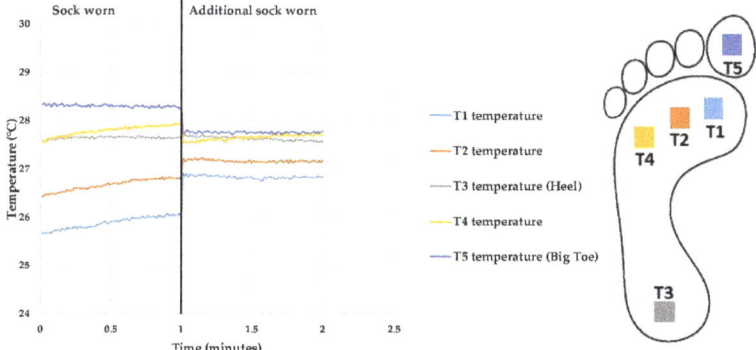

Figure 12. Temperature captured when the temperature sensing sock was worn (presented in the 1st minute) and when an additional sock was worn on top of it (presented in the 2nd minute).

Figure 12 shows that wearing another sock on top of the temperature sensing sock reduced the difference between the temperature captured by the five temperature sensing yarns significantly. Wearing an additional sock provided insulation to external environmental conditions and also enhanced the contact between the skin and the temperature sensing sock.

3.2.6. Discussion

Foot temperature measurements were successfully captured for all the volunteers at five points on the feet. It was observed that the temperatures captured from the first two volunteers were significantly lower than for the last three volunteers. This could be a result of the socks being too large for the feet of the first two volunteers, leading to poor contact between the sock and the skin. The temperature readings from Volunteers 1 and 2 were outside the expected range for foot skin temperature, whereas the temperature readings from the last three volunteers were closer to this value. This proved that the fit of the sock is critical to obtaining meaningful and realistic absolute temperature measurements. This supports the earlier experiments in this paper showing the importance of contact pressure between the temperature sensing yarn and the skin. Therefore, to ensure that the temperature sensing yarn provides precise temperature measurements, the socks should be made to fit the users' feet, guaranteeing good contact between the sock and the skin surface.

It was also observed that critical factors external to the sock, such as the type of footwear worn or movement of the feet, had an effect on the absolute temperature readings. Work by Reddy et al. showed that walking increases the foot temperature and highlighted the importance of creating evidence-based healthcare guidelines for managing diabetic foot ulcerations [34].

The temperature changes recorded in this paper were not necessarily relative. For example, in the case of Volunteer 3, walking induce a −1 °C between points T3 and T4, when they had been in agreement when the foot was static. This would have also had an effect on relative temperature measurements.

For the last three volunteers (Volunteers 3, 4, and 5), the measurements demonstrated that the temperatures recorded for the five temperature sensing yarns varied compared with each other by about ± 2 °C when the socks were worn.

Wearing another sock on top of the temperature sensing sock enhanced the contact between the temperature sensing sock and the skin, acting as an insulating layer between the sock and the external environment. The results showed that the additional sock reduced the recorded temperature differences between points on the foot. The differences between the points were still present when a shoe was worn, which would have provided some degree of thermal insulation. Therefore, the change was likely due to better contact between the foot and the thermistors in the sock. This again highlighted the importance of the fit of the sock on the viability of obtaining foot skin temperature measurements.

Ultimately, highly accurate temperature measurements were only achieved when the sock had a very close fit to the foot. Wearing a shoe and walking also caused dissimilar changes in the temperature measurements from the volunteers (this resulted in an average change of 1.23 °C in the five temperature sensing yarns in Volunteer 3). To obtain highly accurate and statistically relevant measurements, the socks have to be well fitted and calibrated depending on the individual and anatomy of the wearer's shoe. The fit of the sock could be improved by custom knitting socks using scan-to-knit technology [35].

For applications such as non-freezing cold injury detection, where large temperature changes may be observed, and the exposure time to the lower temperatures is regarded important [4], these socks may provide an ideal solution. Currently, no tools are available that are capable of remotely monitoring foot temperature over a prolonged period of time in operational conditions. These socks would provide a powerful tool for the UK military to monitor, record, and analyze the development of non-freezing cold injury in soldiers.

4. Conclusions

Temperature sensing yarns have been successfully used to create a number of temperature sensing garments. The temperature sensing yarns did not impact the textile characteristics of the garments to bend, sheer, and drape. The packing fibers and the warp knitted tube that covered the micro-pod in the temperature sensing yarn ensured that the electronics remained invisible to the wearer and did not affect the feel of the textile garment. Details for producing these garments have been provided along with thorough design considerations. The experiments conducted on hands and a rigid surface showed that contact pressure affects the measurements taken by the temperature sensing yarn due to the deformation of the yarn structure. In terms of the temperature sensing yarn, this is an important result since other temperature sensors that are not textile-based do not deform in this fashion. The experiments highlighted an important limitation in temperature sensing yarn technology; therefore, any fabric made using these yarns would have to be engineered with regard to the contact pressure at the point of the temperature measurement. Once these fabrics are engineered, user trials need to be conducted on a larger scale and this will be completed in the future. This will also be an important design consideration that would have to be taken into account when developing other textile-based temperature sensors.

A trial was conducted with the temperature sensing sock. It was observed that the accuracy of the temperature measurements were heavily dependent on the fit of the sock. Although this might be acceptable for some applications (e.g., non-freezing cold injury), for applications where highly accurate measurements are required, a close fitting of the sock is necessary. This would ensure proper contact between the temperature sensing yarns and the feet of the user. It was also demonstrated that wearing a shoe and an additional sock created a microclimate, which changed the recorded temperature measurements. To better classify the operational limitations of obtaining foot skin temperature measurements with a textile sensor, larger user trials would have to be completed. Identifying the best textile structure to ensure contact between the temperature sensing yarns and the skin is also required. Compression socks, which have already been used for other medical conditions [36], may prove to be an appropriate solution and this will be explored in future work.

Author Contributions: P.L., T.D. conceived and designed the prototypes and the experiments; P.L. performed the experiments and developed the prototype samples; C.O. knitted the prototype garments; P.L. and T.H.-R. analyzed the prototype samples and the experimental data; R.M. contributed specialist expertise in measurement science and interfacing; P.L., T.D. and T.H.-R. wrote the paper.

Funding: This research received funding from the Defence Science and Technology Laboratory through the Centre for Defence Enterprise.

Acknowledgments: Pasindu Lugoda gratefully acknowledges Nottingham Trent University for funding under the Vice Chancellors Bursary Award. The authors would like to thank Richard Arm for fabricating the simulated foot with metal studs. The authors would also like to thank Dorothy Hardy for proofreading the manuscript.

Conflicts of Interest: The authors declare no conflict of interest.

References

1. Hughes-Riley, T.; Lugoda, P.; Dias, T.; Trabi, C.L.; Morris, R.H. A Study of Thermistor Performance within a Textile Structure. *Sensors* **2017**, *17*, 1804. [CrossRef] [PubMed]
2. Chen, W.; Dols, S.; Oetomo, S.B.; Feijs, L. Monitoring Body Temperature of Newborn Infants at Neonatal Intensive Care Units Using Wearable Sensors. In Proceedings of the Fifth International Conference on Body Area Networks, New York, NY, USA, 19–21 June 2010; pp. 188–194.
3. McCallum, L.; Higgins, D. Body temperature is a vital sign and it is important to measure it accurately. This article reviews and compares the various methods available to nurses measuring body temperature. *Nurs. Times* **2012**, *108*, 20–22. [PubMed]
4. Hope, K.; Eglin, C.; Golden, F.; Tipton, M. Sublingual glyceryl trinitrate and the peripheral thermal responses in normal and cold-sensitive individuals. *Microvasc. Res.* **2014**, *91*, 84–89. [CrossRef] [PubMed]
5. Imray, C.; Grieve, A.; Dhillon, S. Caudwell Xtreme Everest Research Group. Cold damage to the extremities: Frostbite and non-freezing cold injuries. *Postgrad. Med. J.* **2009**, *85*, 481–488. [CrossRef] [PubMed]
6. Armstrong, D.G.; Holtz-Neiderer, K.; Wendel, C.; Mohler, M.J.; Kimbriel, H.R.; Lavery, L.A. Skin Temperature Monitoring Reduces the Risk for Diabetic Foot Ulceration in High-risk Patients. *Am. J. Med.* **2007**, *120*, 1042–1046. [CrossRef] [PubMed]
7. Yusuf, S.; Okuwa, M.; Shigeta, Y.; Dai, M.; Iuchi, T.; Rahman, S.; Usman, A.; Kasim, S.; Sugama, J.; Nakatani, T.; et al. Microclimate and development of pressure ulcers and superficial skin changes. *Int. Wound J.* **2015**, *12*, 40–46. [CrossRef] [PubMed]
8. Ring, E.F.J.; Ammer, K. Infrared thermal imaging in medicine. *Physiol. Meas.* **2012**, *33*, R33. [CrossRef] [PubMed]
9. University of Tokyo. Fever Alarm Armband: A Wearable, Printable, Temperature Sensor—ScienceDaily. *ScienceDaily*, 23 February 2015. Available online: http://www.sciencedaily.com/releases/2015/02/150223084343.htm (accessed on 25 February 2015).
10. Nasir Mehmood, A.H. A flexible and low power telemetric sensing and monitoring system for chronic wound diagnostics. *Biomed. Eng. OnLine* **2015**, *14*. [CrossRef]
11. Giansanti, D.; Maccioni, G.; Bernhardt, P. Toward the design of a wearable system for contact thermography in telemedicine. *Telemed. J. E-Health Off. J. Am. Telemed. Assoc.* **2009**, *15*, 290–295. [CrossRef] [PubMed]

12. Klien, T. Spanish Researchers Test Patient Reactions to Wearables for Parkinson Monitoring | EMDT—European Medical Device Technology. *European Medical Device Technology*, 2 November 2015. Available online: http://www.emdt.co.uk/daily-buzz/spanish-researchers-test-patient-reactions-wearables-parkinson-monitoring (accessed on 4 March 2015).
13. Hughes-Riley, T.; Dias, T.; Cork, C. A historical review of the development of electronic textiles. *Fibers* **2018**, *6*, 34. [CrossRef]
14. Husain, M.D.; Kennon, R.; Dias, T. Design and fabrication of Temperature Sensing Fabric. *J. Ind. Text.* **2013**, *44*, 398–417. [CrossRef]
15. Roh, J.S.; Kim, S. All-fabric intelligent temperature regulation system for smart clothing applications. *J. Intell. Mater. Syst. Struct.* **2015**, *27*, 1165–1175. [CrossRef]
16. Ziegler, S.; Frydrysiak, M. Initial Research into the Structure and Working Conditions of Textile Thermocouples. *Fibres Text. East. Eur.* **2009**, *17*, 84–88.
17. Soukup, R.; Hamacek, A.; Mracek, L.; Reboun, J. Textile-based temperature and humidity sensor elements for healthcare applications. In Proceedings of the 2014 37th International Spring Seminar on Electronics Technology, Dresden, Germany, 7–11 May 2014; pp. 407–411.
18. Lavery, L.A.; Higgins, K.R.; Lanctot, D.R.; Constantinides, G.P.; Zamorano, R.G.; Athanasiou, K.A.; Armstrong, D.G.; Agrawal, C.M. Preventing diabetic foot ulcer recurrence in high-risk patients: Use of temperature monitoring as a self-assessment tool. *Diabetes Care* **2007**, *30*, 14–20. [CrossRef] [PubMed]
19. Lavery, L.A.; Higgins, K.R.; Lanctot, D.R.; Constantinides, G.P.; Zamorano, R.G.; Armstrong, D.G.; Athanasiou, K.A.; Agrawal, C.M. Home monitoring of foot skin temperatures to prevent ulceration. *Diabetes Care* **2004**, *27*, 2642–2647. [CrossRef] [PubMed]
20. Cherenack, K.; Zysset, C.; Kinkeldei, T.; Münzenrieder, N.; Tröster, G. Woven Electronic Fibers with Sensing and Display Functions for Smart Textiles. *Adv. Mater.* **2010**, *22*, 5178–5182. [CrossRef] [PubMed]
21. Dias, T.K.; Rathnayake, A. Electronically Functional Yarns. Patent WO 2016038342 A1, 17 March 2016.
22. Dias, T.; Hughes-Riley, T. Electronically Functional Yarns Transform Wearable Device Industry. *R&D Magazine*. 8 January 2017. Available online: https://www.rdmag.com/article/2017/06/electronically-functional-yarns-transform-wearable-device-industry (accessed on 9 September 2017).
23. Lugoda, P.; Dias, T.; Morris, R. Electronic temperature sensing yarn. *J. Multidiscip. Eng. Sci. Stud.* **2015**, *1*, 100–103.
24. Lugoda, P.; Dias, T.; Hughes-Riley, T.; Morris, R. Refinement of temperature sensing yarns. *Proceedings* **2018**, *2*, 123. [CrossRef]
25. Dias, T.; Lugoda, P.; Cork, C.R. Microchip technology used in textile materials. In *High-Performance Apparel*, 1st ed.; McLoughlin, J., Sabir, T., Eds.; Woodhead Publishing Ltd.: Cambridge, UK, 2017.
26. Hardy, D.A.; Anastasopoulos, I.; Nashed, M.N.; Oliveira, C.; Hughes-Riley, T.; Komolafe, A.; Tudor, J.; Torah, R.; Beeby, S.; Dias, T. An Automated Process for Inclusion of Package Dies and Circuitry within a Textile Yarn. In Proceedings of the Design, Test, Integration & Packaging of MEMS/MOEMS, Rome, Italy, 22–25 May 2018.
27. Gould, P. Textiles gain intelligence. *Mater. Today* **2003**, *6*, 38–43. [CrossRef]
28. Munro, B.J.; Campbell, T.E.; Wallace, G.G.; Steele, J.R. The intelligent knee sleeve: A wearable biofeedback device. *Sens. Actuators B Chem.* **2008**, *131*, 541–547. [CrossRef]
29. Stuart, I.M. Capstan equation for strings with rigidity. *Br. J. Appl. Phys.* **1961**, *12*, 559. [CrossRef]
30. ASTM International. *Manual on the Use of Thermocouples in Temperature Measurement*; ASTM International: West Conshohocken, PA, USA, 1993; p. 175. Available online: https://drive.google.com/file/d/0B7YMfVp35H91NTU0Y2QwZWYtYjA3Yy00NTJiLWJiYzEtZjQ3ZjljNjhhNWIy/view?hl=en (accessed on 3 March 2015).
31. KRUSS. Laplace Pressure. 2018. Available online: https://www.kruss-scientific.com/services/education-theory/glossary/laplace-pressure/ (accessed on 23 February 2018).
32. Psikuta, A.; Niedermann, R.; Rossi, R.M. Effect of ambient temperature and attachment method on surface temperature measurements. *Int. J. Biometeorol.* **2013**, *58*, 877–885. [CrossRef] [PubMed]
33. Nardin, R.A.; Fogerson, P.M.; Nie, R.; Rutkove, S.B. Foot temperature in healthy individuals: Effects of ambient temperature and age. *J. Am. Podiatr. Med. Assoc.* **2010**, *100*, 258–264. [CrossRef] [PubMed]

34. Reddy, P.N.; Cooper, G.; Weightman, A.; Hodson-Tole, E.; Reeves, N.D. Walking cadence affects rate of plantar foot temperature change but not final temperature in younger and older adults. *Gait Posture* **2017**, *52*, 272–279. [CrossRef] [PubMed]
35. Power, E.J. Chapter 12—Yarn to Fabric: Knitting. In *Textiles and Fashion*; Sinclair, R., Ed.; Woodhead Publishing: Cambridge, UK, 2015; pp. 289–305.
36. Raju, S.; Kathryn, H.; Neglen, P. Use of Compression Stockings in Chronic Venous Disease: Patient Compliance and Efficacy. Presented at the Nineteenth American Venous Forum Meeting, San Diego, CA, USA, 14–17 February 2007.

© 2018 by the authors. Licensee MDPI, Basel, Switzerland. This article is an open access article distributed under the terms and conditions of the Creative Commons Attribution (CC BY) license (http://creativecommons.org/licenses/by/4.0/).

MDPI
St. Alban-Anlage 66
4052 Basel
Switzerland
Tel. +41 61 683 77 34
Fax +41 61 302 89 18
www.mdpi.com

Fibers Editorial Office
E-mail: fibers@mdpi.com
www.mdpi.com/journal/fibers

www.ingramcontent.com/pod-product-compliance
Lightning Source LLC
LaVergne TN
LVHW072000080526
838202LV00064B/6807